Smart Sensors and Systems

Yongpan Liu • Youn-Long Lin • Chong-Min Kyung
Hiroto Yasuura

Editors

Smart Sensors and Systems

Technology Advancement and Application
Demonstrations

 Springer

Editors
Yongpan Liu
Circuits and Systems Division
Tsinghua University Circuits
and Systems Division
Beijing, China

Chong-Min Kyung
#310 IT Convergence Building (N1)
Center for Integrated Smart Sensors
Yuseong-gu, Daejeon, Korea
(Democratic People's Republic of)

Youn-Long Lin
National Tsing Hua University
Hsichu, Taiwan

Hiroto Yasuura
Kyushu University
Fukuoka Shi, Japan

ISBN 978-3-030-42236-3   ISBN 978-3-030-42234-9   (eBook)
https://doi.org/10.1007/978-3-030-42234-9

This Springer imprint is published by the registered company Springer Nature Switzerland AG.
The registered company address is: Gewerbestrasse 11, 6330 Cham, Switzerland

Preface

Internet of Things (IoTs) is becoming the key technology for smart social information systems. To realize the smart society by IoTs, effective sensing technologies of various aspects of physical world are required. The huge sensing data should be intelligently processed and integrated by smart sensor systems. This book shows essential issues of the IoTs and smart sensor technologies in various aspects, from fundamental devices to actual social applications.

Following previous three editions, this book brings together multidisciplinary sensor technology from biological, optical, chemical, and electrical domains. The research field is expanding to various application areas, and new researches are being explored. This book presents up-to-date approaches of sensor devices and smart sensor systems in real-world applications including biomedical, video, and fundamental IoT techniques.

Chapters 3, 4, and 9 illustrate biosensor techniques, including optogenetics-based implantable neural interfaces, plasmonic nanostructures for biomarker sensing, and bio-imaging and PDMS film process for thin and flexible sensors. Specifically, (1) Chap. 3 introduces optogenetics, one of the most powerful research tools to selectively control or modulate the activity of specific types of neurons. As an alternative light source in place of laser, two different types of LED-based implantable neural interfaces are compared in terms of efficiency, fabrication complexity, multifunctionality, and biosafety. (2) Chapter 4 presents recent progress in the fabrication of practical plasmonic nanostructures for surface-enhanced Raman spectroscopy (SERS) applications. A nano-transfer printing of sub-20 nm metallic nanostructures is illustrated for cost-effective and reproducible plasmonic nanostructures, which is promising for biomedical applications such as biomarker sensing and bio-imaging. (3) Chapter 9 proposes a low-cost processing technique for microstructuring PDMS film. Pressure and strain sensors are fabricated with these PDMS films and integrated in wearable systems for monitoring diverse human physiological signals and body motions, including wrist pulse monitoring, static and dynamic foot pressure monitoring, and neck posture monitoring.

Chapters 1, 2, and 5 present better performance and energy efficiency techniques for video sensor applications, consisting of a Poisson Mixture Model (PMM)-based background subtraction, vision-based and Bayesian filtering motion estimation, and simultaneous localization and mapping algorithms for visual tracking. Specifically, (1) Chap. 1 presents a new background subtraction scheme for camera systems based on a PMM for modeling dynamic backgrounds, and it achieves a significant speed-up compared to FPGA implementations, making it especially suitable for embedded applications. (2) Chapter 2 improves motion estimation approaches by a novel visual simultaneous localization and mapping (SLAM) approach in which both vision-based motion estimation and Bayesian filtering are combined to reduce the estimation errors of the estimated path. (3) Chapter 5 presents basics and recent advances in visual tracking for augmented reality, computer vision, and robotics applications. Specifically, it focuses on visual simultaneous localization and mapping (vSLAM) algorithms that allow both camera pose estimation and 3D model generation in unprepared environments.

Chapters 6, 7, 8, and 10 discuss novel IoT techniques including wireless backhaul network techniques, vehicle detection using sidewalk microphones, high-performance algorithm-architecture co-design and nonvolatile memory-based IoT sensor nodes. Specifically, (1) Chap. 6 proposes a wireless backhaul network in which access points (APs) are linked wirelessly with the capability of relaying packets, which realizes a wide Wi-Fi coverage area without installing a huge number of access cables. Four example applications are described: a ubiquitous camera network, a Wi-Fi tag tracking system, a criminal fishing system, and a networked vehicle. (2) Vehicle detection is one of the fundamental tasks in the ITS (intelligent transportation system). Chapter 7 presents a new approach of vehicle sensing using sidewalk microphones for vehicle detection. (3) Chapter 8 innovates discussions on Smart System-on-Chip design, in expediting the field of signal and information processing systems into futuristic new era of the IoTs and high-performance computing based on algorithm/architecture co-design. (4) Chapter 10 discusses architecting PCM, especially MLC PCM, as main memory for IoT devices to replace conventional DRAM deployment. However, PCM/MLC PCM suffers from long write latency and large write energy; this work proposes a write mode aware loop tiling approach to effectively reduce the lifetime of write instances and maximize the number of efficient fast writes in loops.

Beijing, China   Yongpan Liu
Hsichu, Taiwan   Youn-Long Lin
Daejeon, Korea   Chong-Min Kyung
Fukuoka Shi, Japan   Hiroto Yasuura

Contents

Poisson Mixture Model for High Speed and Low-Power Background Subtraction

Muhammad Umar Karim Khan and Chong-Min Kyung

1 Introduction

Internet of Things (IoT) has been driving research in industry as well as academia for the past decade. An IoT system presents numerous challenges to designers of different expertise. One of the major tasks of designing an IoT system is to develop the sensors at the front-end of the system. These sensors are required to perform their respective tasks in real-time with limited energy and area. Inefficient implementation of sensors in IoT clusters with battery-operated nodes will result in limited operational time, as high energy consumption will quickly drain the battery. Although energy-efficient communication in IoT has been thoroughly researched, relatively little progress has been made in the design of energy-efficient sensors.

Vision sensors in the form of smart cameras are expected to be a core part of the IoT. This is because many real-life applications depend on visual understanding of a scene. Designing a smart camera for IoT that infers from the scene is a challenging proposition. First, computer vision algorithms have high computational complexity, making them inefficient for IoT. Second, accuracy is generally compromised in hardware implementations of computer vision algorithms for achieving higher speed, lower power, or smaller area.

Background subtraction (BS) is a core computer vision algorithm to segment moving objects from the dynamic or static background. BS algorithms aim to model the background that is robust against lighting changes, camera jitter, and dynamic

M. U. K. Khan (✉)
KAIST, Daejeon, South Korea
e-mail: umar@kaist.ac.kr

C.-M. Kyung
#310 IT Convergence Building (N1), Center for Integrated Smart Sensors, Yuseong-gu, Daejeon, Korea (Democratic People's Republic of)
e-mail: kyung@kaist.ac.kr

© Springer Nature Switzerland AG 2020
Y. Liu et al. (eds.), *Smart Sensors and Systems*,
https://doi.org/10.1007/978-3-030-42234-9_1

backgrounds [1]. Generally, BS provides a region of interest (the moving object) in the frame of a given video, which is either used to trigger an alarm or further analyzed to understand the scene. Computational gain is achieved by only analyzing the moving objects in the scene rather than the whole frame.

Researchers have proposed many BS schemes in the past. Almost all of these schemes target high accuracy of BS. Such an approach is useful in some applications but not for smart cameras. The reason is that the algorithms that solely target high accuracy are computationally complex, resulting in high power consumption, delay, and area. On the other hand, some researchers have proposed dedicated implementations of BS, which provide relatively high speed and low-power consumption. However, these implementations generally degrade the accuracy of the algorithms.

Previously, we have described the use of BS in surveillance systems, its implications and proposed a novel method. Khan et al. [2] presents a strategy for obtaining the ideal learning rate of GMM-based BS to minimize energy consumption. A dual-frame rate system was proposed in [3], which allows efficient use of memory in a blackbox system. We proposed an accurate, fast, and low-power BS scheme in [4].

Shot noise is the most dominant source of noise in camera system. Shot noise is modeled by a Poisson distribution. Therefore, in this paper, we use a Poisson distribution under the shot noise assumption to model the background pixel intensity. In fact, we use a Poisson mixture model (PMM) to model dynamic background. We use a relatively stable approach for online approximation of the parameters of the distribution. Resultantly, the proposed method provides competitive performance compared to common BS schemes.

The rest of the chapter is structured as follows. Section 2 describes some efficient implementations of BS from the literature. In Sect. 3, we present a brief review of EBSCam for BS. Section 4 describes the proposed method of BS. Experimental results are discussed in Sect. 5 and Sect. 6 concludes the chapter.

2 Previous Work

This section is divided into two parts. In the first part, we describe numerous BS algorithms. In the second part, efficient implementations of BS algorithms are presented.

2.1 Background Subtraction Algorithms

Previously, numerous surveys have been performed on BS schemes [5–7]. These BS schemes can be classified into region-based and pixel-based categories.

Region-based schemes make use of the fact that background pixels of the same object have similar intensities and variations over time. In [8], authors divide a frame into $N \times N$ blocks and each block is processed separately. Samples of the N^2 vectors are then used to train a principal component analysis (PCA) model. The PCA model is used for foreground classification. A similar technique is described in [9]. In [10], independent component analysis (ICA) of images from a training sequence is used for foreground classification. A hierarchical scheme for region-based BS is presented in [11]. In the first step, a block in the image is classified as background and the block is updated in the second step.

Pixel-based BS schemes have attracted more attention due to their simpler implementations. In these schemes, the background model is maintained for each pixel in the frame. The simplest of these methods are classified as frame differencing, where the background is modeled by the previous frame [12] or by the (weighted) average of the previous frames. Authors in [13] use the running average of most recent frames so that old information is discarded from the background model. This scheme requires storing some of the previous frames; thereby, larger memory space is consumed. In [14] and [15], a univariate Gaussian distribution is associated with each pixel in the frame, i.e., pixel intensities are assumed to be of Gaussian distribution and the parameters of the distribution (mean and variance) for each pixel are updated with every incoming frame. The mean and the variance of the background model for each pixel are then used to classify foreground pixels from the background.

Sigma-delta (Σ-Δ)-based BS [16, 17] is another scheme which is popular in embedded applications [18, 19]. Inspired from analog-to-digital converters, these methods use simple non-recursive approximates of the background image. The background pixel intensity is incremented, decremented, or unchanged if the pixel intensity is considered greater than, less than, or similar to the background pixel intensity, respectively.

Kernel density estimation (KDE) schemes [20, 21] accumulate the histogram of pixels separately in a given scene to model the background. Although claimed to be non-parametric, the kernel bandwidth of KDE-based schemes needs to be decided in advance. Using a small bandwidth produces rough density estimates whereas a large bandwidth produces over-smoothed ones.

Perhaps the most popular BS methods are the ones based on Gaussians mixture models (GMM). These methods, first introduced in [22], assume that background pixels follow a Gaussian distribution and model a background pixel with multiple Gaussian distributions to include multiple colors of the background. Numerous improvements have been suggested to improve foreground classification [23] as well as speed [24–27] of GMM-based methods. Notable variants of the original work are [28] and [29]. In [28], an adaptive learning rate is used to update the model. In [29], which is usually referred to as extended GMM or EGMM, author uses the Dirichlet prior with some of the update equations to determine the sufficient number of distributions to be associated with each pixel. The Flux Tensor with Split Gaussians (FTSG) scheme [30] uses separate models for the background and foreground. The method develops a tensor based on spatial and temporal information, and uses the tensor for BS.

Another pixel-based method for BS is the codebook scheme [31] and [32], which assigns a code word to each background pixel. A code word, extracted from a codebook, indicates the long-term background motion associated with a pixel. This method requires offline training and cannot add new code words to the codebook at runtime.

ViBe [33] is a technique enabling BS at a very high speed. The background model for each pixel includes some of the previous pixels at the given pixel location as well as from the neighboring locations. A pixel identified as background replaces one of the randomly selected background pixels for the corresponding and the neighboring pixel locations. The rate of update is controlled by a fixed parameter called the sampling factor. Despite its advantages, the BS performance of ViBe is unsatisfactory in challenging scenarios such as dark backgrounds, shadows, and under frequent background changes [34].

PBAS [35] is another BS scheme that only maintains background samples in the background model. PBAS has a similar set of parameters as ViBe. It operates on the three independent color channels separately, and combines their results for foreground classification. Another sample based scheme is SACON [36]. This method computes a sample consensus by considering a large number of previous pixels, similar to ViBe. Authors assign time out map values to pixels and objects, which indicate for how long a pixel has been observed in consecutive frames. These values are used to add static foreground objects to the background.

Recently, several BS schemes based on human perception have been proposed. In [37], authors assumed that human vision does not visualize the scene globally but is rather focused on key-points in a given scene. Their proposed method uses key-point detection and matching for maintaining the background model. Saliency has been used in [38] for developing a BS technique. In [34], authors consider how the human visual system perceives noticeable intensity deviation from the background. Authors in [39] present a method of BS for cell-phone cameras under view angle changes.

2.2 Hardware Implementation of Background Subtraction Schemes

Some implementations of BS algorithms for constrained environments have been proposed in the past. The BS algorithm presented in [40] has been implemented on a Spartan 3 FPGA in [41]. Details of the implementation are missing in their work. Furthermore, [40] assumes the background to be static, which is impractical. In [42], authors present a modification to the method proposed in [43]. They have achieved significant gains in memory consumption and execution time; however, they have not presented the BS performance results over a complete dataset. Authors in [44] implement the algorithm presented in [45] on the Spartan 3 FPGA. The implemented algorithm, however, is non-adaptive and applicable to static backgrounds only. Furthermore, the algorithm of [45] lacks quantitative evaluation. An implementation of single Gaussian-based BS on a Digilent FPGA is given in [46]. As [45], the implemented algorithm cannot model dynamic backgrounds. Another implementation of a BS scheme for static backgrounds, more specifically the selective running Gaussian filter-based BS, has been performed on a Virtex 4

FPGA in [47]. In [48], authors present a modified multi-modal Σ-Δ BS scheme. They have achieved very high speed with their implementation on a Spartan II FPGA. Like other methods discussed above, they have not evaluated the BS performance of their method on a standard dataset.

Many researchers have implemented comparatively better performing BS schemes as well. A SoC implementation of GMM is presented in [49], which consumes 421 mW. In [50], an FPGA implementation of GMM is presented, which is faster and requires less energy compared to previous implementations. The authors maneuver the update equations to simplify the hardware implementations. Other implementations of GMM include [51] and [52]. An implementation of the codebook algorithm for BS is presented in [53]. Similarly, FPGA implementations of ViBe and PBAS algorithms have presented in [54] and [55], respectively.

It should be noted that the above implementations do not exactly implement the original algorithms but use a different set of parameters or post-processing to adapt their methods for better hardware implementations; therefore, the BS performance of these implementations is expected to be different from the original algorithms.

3 Review of EBSCam

EBSCam is a BS scheme proposed in [4]. It uses a background model that is robust against the effect of noise. In this work, we use EBSCam to estimate the parameters of the PMM distribution for every pixel.

It is shown in [4] that the noise in the input samples results in the background model of each pixel to fluctuate. This results in BS scheme making classification errors, i.e., identifying background pixels as foreground and vice versa. These errors are typically defined in terms of false positive and false negatives. A false positive is said to occur if a pixel belonging to the background is identified to be part of a moving object. Similarly, a false negative is said to occur if a pixel belonging to the moving object is identified as to be part of the background. In [4], the probability of false positives and false negatives with GMM-based BS is shown to be

$$P[FP] = 1 - \text{erf}\left(\frac{\sqrt{T}\sigma_{BG}}{\sqrt{2\left(\sigma_{BG}^2 + s_{\mu,k}^2(\alpha_k, \sigma_{BG}) + \psi + \sqrt{T}s_{\sigma,k}^2(\alpha_k, \sigma_{BG})\right)}}\right) \tag{1}$$

and

$$P[FN] = 1 - \text{erf}\left(\frac{\text{E}[I_{FG}] - \left(\text{E}[I_{BG}] + \sqrt{T}\sigma_{BG}\right)}{\sqrt{2\left(\sigma_{FG}^2 + s_{\mu,k}^2(\alpha_k, \sigma_{BG}) + \sqrt{T}s_{\sigma,k}^2(\alpha_k, \sigma_{BG})\right)}}\right), \tag{2}$$

respectively. Note that here $s^2_{\mu,k}$ and $s^2_{\sigma,k}$ denote the variance in the estimated mean and standard deviation parameters of GMM, respectively. Kindly, refer to [4] for the definition of the rest of the symbols as these are not relevant here. From the above equations it is seen that the variance in the estimated parameters increases the error probabilities. EBSCam mitigates this variance in the estimated parameters to reduce errors in BS.

In EBSCam, the intensity of the background of each pixel is limited to have K different intensities at maximum. In other words, the background can have K different layers to allow the method to estimate dynamic backgrounds. For every pixel i, a set E_i of K elements is formed. Each element of E_i represents a single layer of the background.

The pixel intensity at i-th pixel is compared against the elements of the set E_i to populate E_i. If the pixel intensity differs from all the elements of the set E_i by more than a constant D, then it is stored as a new element in the set E_i. Let us define

$$R_{i,t} = \cup^K_{j=1}[E^{(j)}_{i,t-1} - D, E^{(j)}_{i,t-1} + D], \tag{3}$$

where $E^{(j)}_{i,t-1}$ is the j-th element of E_i at time $t - 1$ and D is a global threshold, then the sampling frame (if the input intensity belongs to $S_{i,t}$ then it is included in E_i) at time t is defined as

$$S_{i,t} = I \setminus R_{i,t}. \tag{4}$$

Here I is the set of all possible values of background pixels and $\setminus$ denotes set subtraction.

It is seen from the above equations that the background model only changes when the input intensity differs from all the elements of E_i by more than D, otherwise, the background model does not fluctuate. In other words, triggering the update of the background model is thresholded by a step-size of D in EBSCam. Furthermore, pixel intensities can be used for estimating the background intensity as under the assumption of a normal distribution the mean and the mode are the same. In other words, the most frequently observed value of the background intensity is likely to be very close to the mean of the intensity, i.e.,

$$\arg\max_{I_{BG}} f_{I_{BG}}(I_{BG}) = \mathrm{E}[I_{BG}], \tag{5}$$

where $f_{I_{BG}}$ is the probability density function (PDF) of the background intensity.

The background model in EBSCam is not limited to the estimates E_i but also the credence of individual estimates, which is stored in a set C_i. The credence gives the confidence in each estimate. In [4], it is shown that the credence should be incremented if an estimate is observed and should be decremented if the respective estimate is not observed. More precisely,

$$C^{(j)}_{i,t} = C^{(j)}_{i,t-1} + 1 \text{ if } I_{i,t} \in [E^{(j)}_{i,t-1} - D, E^{(j)}_{i,t-1} + D] \tag{6}$$

and

$$C_{i,t}^{(j)} = C_{i,t-1}^{(j)} - 1 \text{ if } (I_{i,t} \in [E_{i,t-1}^{(j)} - D, E_{i,t-1}^{(j)} + D]' \wedge C_{i,t-1}^{(j)} > C_{th}), \qquad (7)$$

where C_{th} is a constant.

The background model in EBSCam is updated blindly. In blind updates, the input intensity at a pixel updates the background model regardless of the pixel intensity being part of the background or foreground. On the other hand, in non-blind updates only the background pixel intensities are used to update the background model. Whenever a pixel intensity that differs from all the elements of E_i by more than D is observed then it is included as a new estimate. The new estimate has a credence value of zero. Note that the new estimate replaces the estimate about which we are least confident, i.e., the estimate with the minimum credence.

The background model is used to classify pixels to either belonging to the background model (background pixels) or the foreground. The decision to classify a pixel to background or foreground is straightforward. First, the set of estimates about which we are confident enough are defined as

$$B_{i,t} = \cup_{j=1}^{K}(E_{i,t-1}^{(j)}|C_{i,t-1}^{(j)} > C_{th}), \qquad (8)$$

i.e., if the credence of an estimate is greater than a threshold then we can consider the estimate to be valid. The pixel intensity is compared against the valid background estimates. If the pixel intensity matches the valid background estimates, then it is considered as part of the background, otherwise, it is considered to be part of the foreground. The foreground mask at a pixel i is obtained as

$$F_{i,t} = \begin{cases} 0 \text{ if } I_{l,t} \subset [B_{i,t}^{(m)} \quad D, B_{i,t}^{(m)} \mid D] \text{ over any } m \\ 1 \qquad\qquad\qquad\qquad\qquad \text{otherwise} \end{cases} \qquad (9)$$

where $B_{i,t}^{(m)}$ is the m-th element of $B_{i,t}$.

4   EBSCam with Poisson Mixture Model

In EBSCam, a fixed threshold is used for distinguishing the foreground pixels from the background. Although such an approach has been used by numerous authors, such as [33], it lacks theoretical foundation. The variation in the background pixel intensities varies spatially and temporally over video frames; therefore, an adaptive threshold should be used for distinguishing the foreground pixels from the background pixels.

Generally, the Poisson distribution is used to model the shot noise of image sensors. The probability that m photons have been absorbed at a pixel i is given by

$$(X_{i,t} = m) \sim g(m; \lambda_i) = \frac{\lambda_i^m e^{-\lambda_i}}{m!}, \tag{10}$$

where λ_i is a parameter denoting both the mean and the variance of the distribution. Assuming that the number of photons are such that the relationship between the observed intensity and the number of photons is linear, the observed intensity can be given by

$$I_{i,t} = a X_{i,t}, \tag{11}$$

where a is the gain factor. Thus, the mean and the variance of the observed intensity are given by

$$\mu_{i,t} = a \lambda_i \tag{12}$$

and

$$\sigma_{i,t}^2 = a^2 \lambda_i = a \mu_{i,t}, \tag{13}$$

respectively.

The Poisson distribution can be used to model the noise or the variance of the background pixel intensities. In order to deal with dynamic backgrounds, we propose using a Poisson mixture model (PMM) for modeling the background intensities. Each subpopulation of the mixture model is representative of a layer of the background. Thus, the background pixel intensity can be modeled as

$$(I_{BG,i,t} = m) \sim \sum_{k=1}^{K} \psi_i^{(k)} g(m; \lambda_i^{(k)}). \tag{14}$$

The parameters of the distribution can be estimated as

$$\lambda_i^{(k)} = \frac{\sum_{t=1}^{T} 1\{I_{BG,i,t} \in k\} I_{BG,i,t}}{\sum_{t=1}^{T} 1\{I_{BG,i,t} \in k\}} \tag{15}$$

and

$$\psi_i^{(k)} = \frac{\sum_{t=1}^{T} 1\{I_{BG,i,t} \in k\}}{T}, \tag{16}$$

where T is the total number of frames. Generally, expectation maximization algorithm is used to estimate the above parameters.

The approach in (15) and (16) for estimating the parameters of the distribution of the background pixel intensities is not feasible for two reasons. First, it requires maintaining all the frames of the video. Second, EM is an iterative procedure and is constrained by speed requirements. Here, we propose using EBSCam for online approximation to estimate the parameters of the PMM. Since the mean and the mode of the Poisson distribution are the same, we can write

$$\lambda_{i,t}^{(k)} = E_{i,t}^{(k)}. \tag{17}$$

Rather than using the normalized values of $\psi_i^{(k)}$ for which $\sum_{k=1}^{K} \psi_i^{(k)} = 1$, we can use non-normalized values by replacing $\psi_i^{(k)}$ of (16) by

$$\phi_i^{(k)} = \sum_{n=1}^{T} 1\{I_{BG,i,n} \in k\}. \tag{18}$$

This approximation is plausible for the reason that the $\psi_i^{(k)}$ values are used for comparison between subpopulations. Since overall k division by T takes place as shown in (16), therefore, the scaling of $\psi_i^{(k)}$ by T will not have an effect on the result of comparison over all k. From (18),

$$\phi_{i,t}^{(k)} = \sum_{n=1}^{t} 1\{I_{BG,i,n} \in k\} = \phi_{i,t-1}^{(k)} + 1\{I_{BG,i,t} \in k\}. \tag{19}$$

In detail, the approximate weight of the k-th subpopulation is incremented if the input intensity matches the k-th subpopulation. Similarly, to get rid of old, unobserved background layers we decrement $\phi_{i,t}^{(k)}$ if $I_{BG,i,t}$ does not belong to the k-th subpopulation. In case a new background layer needs to be stored, the least observed background layer is to be removed. Based on this, the approximate weight $\phi_{i,t}^{(k)}$ can be replaced by $C_{i,t}^{(k)}$.

The decision whether the input intensity matches an estimate is modified as

$$\left(I_{i,t} - E_{i,t-1}^{(j)}\right)^2 < c^2 \left(\sigma_{i,t-1}^{(j)}\right)^2 \tag{20}$$

or

$$\left(I_{i,t} - E_{i,t-1}^{(j)}\right)^2 < c^2 E_{i,t-1}^{(j)} \tag{21}$$

or

$$\left|I_{i,t} - E_{i,t-1}^{(j)}\right|^2 < c\sqrt{E_{i,t-1}^{(j)}}, \tag{22}$$

where c is a constant. Thus, an adaptive threshold is used to distinguish foreground intensities from the background. The threshold is determined by the mean or the variance of PMM, which can be approximated by E_i.

Algorithm 1 EBSCam-PMM

1: **INPUT** : $I_t \leftarrow$ video frame at time t, $P \leftarrow$ set of all pixels, $K \leftarrow$ cardinality of the estimates
2: **OUTPUT** : $F_t \leftarrow$ foreground mask at time t
3: All elements of $\boldsymbol{E}_i$ and $\boldsymbol{C}_i$ are set to zero for the first frame
4: **for** $i = 1$ to P **do**
5:    $F_{i,t} = 1$;
6:    $match = 0$;
7:    **for** $j = 1$ to K **do**
8:       $diff = |I_{i,t} - E_{i,t}^{(j)}|$;
9:       **if** $diff < c\sqrt{E_{i,t}^{(j)}}$ **then**   ▷ Estimate is observed
10:          $C_{i,t}^{(j)} = C_{i,t-1}^{(j)} + 1$;
11:          $match = 1$;
12:          **if** $C_{i,t-1}^{(j)} > C_{th}$ **then**
13:             $F_{i,t} = 0$;
14:          **end if**
15:       **else**   ▷ Estimate is not observed
16:          **if** $C_{i,t-1}^{(j)} > C_{th}$ **then**
17:             $C_{i,t}^{(j)} = C_{i,t-1}^{(j)} - 1$;
18:          **end if**
19:       **end if**
20:    **end for**
21:    **if** $match == 0$ **then**
22:       Find m such that $C_{i,t}^{(m)} < C_{i,t}^{(j)}$ over $j = 1 : K$
23:       $E_{i,t}^{(m)} = I_{i,t}$;   ▷ Initializing the estimate
24:       $C_{i,t}^{(m)} = 0$;
25:    **end if**
26: **end for**

The update of $\boldsymbol{E}_i$ and foreground classification is performed similar to EBSCam. Since, the proposed method uses EBSCam to estimate the distribution parameters of PMM, we term the proposed scheme as EBSCam-PMM. A step-wise procedure is shown in Algorithm 1.

5 Parameter Selection

In EBCam-PMM, there are two parameters namely c and K. If a too large value of K is used, then it will cover more than the background model, resulting in false negatives. Similarly, a too small value of K will result in false positives, as background intensities will not be fully covered by the estimated mixture model. For c we propose using a value of 2, i.e., the intensity at a pixel is said to match a background estimate if it is within two standard deviations of the estimate. We choose $c = 2$ for a couple of reasons. First, two standard deviations sufficiently cover the distribution. Second, assuming integer values of the input intensity, we can write the condition in (22) as

$$\left| I_{i,t} - E_{i,t-1}^{(j)} \right| < \left\lfloor c\sqrt{E_{i,t-1}^{(j)}} \right\rfloor. \tag{23}$$

With a non-negative integer c, the above can be written as

$$\left| I_{i,t} - E_{i,t-1}^{(j)} \right| < c \left\lfloor \sqrt{E_{i,t-1}^{(j)}} \right\rfloor. \tag{24}$$

By using (24), we can replace the complicated square-root operation by simple LUTs. For example, for all values of $E_{i,t}^{(j)}$ between 170 and 224, the value of $\left\lfloor \sqrt{E_{i,t-1}^{(j)}} \right\rfloor$ is 14. Generally, for an n-bit wide input intensity, we require only $\lfloor 2^{\frac{n}{2}} \rfloor$ if-else conditions to compute $\left\lfloor \sqrt{E_{i,t-1}^{(j)}} \right\rfloor$.

We applied EBSCam-PMM to the dataset in [56] with different values of K. In Fig. 1, we show the percent of wrong classification (PWC) defined as

$$\text{PWC} = \frac{FP + FN}{TP + TN + FP + FN} \times 100. \tag{25}$$

Here TP are true positives and TN are true negatives. With EBSCam-PMM, optimal performance is achieved with $K = 5$ as seen in Fig. 1.

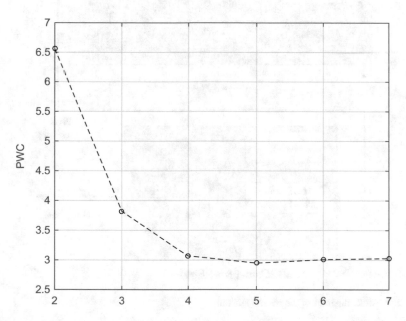

Fig. 1 The effect of changing K on background subtraction performance of EBSCam-PMM over CDNET-2014 dataset

6 Hardware Implementation of EBSCam-PMM

In this section, a dedicated implementation of EBSCam-PMM is described as dedicated implementations can attain much higher processing speed and much lower energy compared to a general purpose implementation.

The abstract diagram of the overall system is shown in Fig. 2. The scene is captured by the sensor array. Afterwards, the intensity values of the scene are passed to the image signal processor (ISP). The ISP performs multiple image processing tasks, which include providing the luminosity values which are used in BS. Note that this front-end configuration is not fixed and can be replaced by any system which provides luminosity values of the scene.

The EBSCam-PMM engine performs the task of identifying the foreground pixels from the background. The EBSCam-PMM engine is composed of the memory unit and the EBSCam-PMM circuit. The memory unit maintains the background model, which is used by the EBSCam-PMM circuit to classify a given pixel into foreground or background.

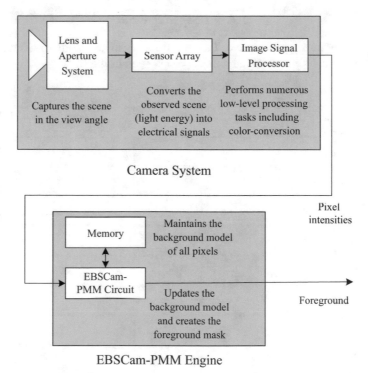

Fig. 2 Abstract diagram of the overall system

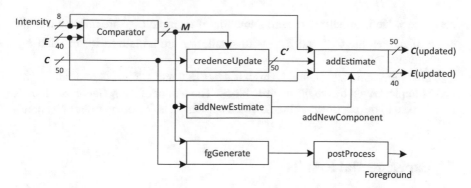

Fig. 3 Block diagram of EBSCam-PMM circuit. Bit-widths are based on FPGA implementation (Sect. 7)

The block diagram of the BS circuit is shown in Fig. 3. This implementation uses 8 and 10 bits for pixel intensities and $C_i^{(j)}$ for all j, respectively. In the graphical representation of each of the constituent modules of EBSCam-PMM circuit, we have excluded the pixel index i as the same circuit is used for all pixels. Similarly, we have excluded the time index from the notation in this section as it can be directly derived from the time index of input intensity. Also, we have not included clock and control signals in the figures to emphasize the main data-flow of the system.

The *comparator* module in the BS circuit compares all the elements of E_i with $I_{i,t}$ in parallel, and generates a K-bit wide output M. $M^{(j)} = 1$ indicates that the pixel has matched $E_i^{(j)}$ and vice versa, which is determined by comparing $|I_{i,t} - E_i^{(j)}|$ with $c \left\lfloor \sqrt{E_{(i,t-1)}^{(j)}} \right\rfloor$. The rest of the hardware is the same for EBSCam and EBSCam-PMM. Note that multiple $M^{(j)}$ can be high at the same instant.

A new estimate needs to be added to the background model if $M^{(j)} = 0$ for $j = 1$ to K. The *addNewEstimate* module checks this condition by performing a logical-NOR of all the bits of M.

Next, the *credenceUpdate* module updates $C_i^{(j)}$ values based on $M^{(j)}$, i.e., *credenceUpdate* module is an implementation of (6) and (7).

The *addEstimate* module is used to add a new estimate to the background model. A new estimate is added to the background model of a pixel if the output of *addNewEstimate* module is high. The module is further subdivided into two submodules.

The *replaceRequired* submodule determines the index of the estimate which needs to be replaced, and the *replaceEstimate* submodule generates the updated estimates and credence values. If required, the *addEstimate* module replaces the estimate with minimum credence value by the intensity of the pixel. Also, the credence value is initialized to zero. Note that if multiple $C_i^{(j)}$ are minimum at the same time, then the $C_i^{(j)}$ and $E_i^{(j)}$ with smallest j are initialized to zero and $I_{i,t}$, respectively.

The foreground pixel should be high if $M^{(j)} = 0$ and $C_i^{(j)} > C_{th}$ for all j. This task is implemented by the *fgGenerate* block. The output of the *fgGenerate* block is then passed to the *postProcess* block which applies a 7×7 median filter to its input.

7   Experimental Results

To analyze the performance of EBSCam-PMM, we present results of applying EBSCam-PMM to standard datasets. Also, in this section we discuss the FPGA implementation and results of EBSCam-PMM. The performance of the proposed method is compared against some state-of-the-art implementations as well.

7.1   Background Subtraction Performance

To analyze the accuracy of BS under different scene conditions, we have applied EBSCam-PMM to the CDNET-2014 dataset, which is the most thorough BS evaluation dataset available online. CDNET-2014 [56] is an extensive dataset of real-life videos. It includes 53 video sequences divided into 11 categories.

We have used a fixed set of parameters for evaluation of our method. In practice, the value of C_{th} should be varied with the frame rate of the video. However, here we have used a fixed value of C_{th} over the whole dataset. We have used a 7×7 median filter as a post-process. The PWC metric has been used to evaluate the performance, as it is a commonly used performance metric to evaluate and compare binary classifiers such as BS.

In Table 1, we present and compare the performance of EBSCam-PMM with GMM [22], EGMM [29], KDE [57], ViBe [33], PBAS [35], and EBSCam [4]. From Table 1, it is seen that EBSCam-PMM shows improved performance compared to GMM, EGMM, KDE, and ViBe and is only next to PBAS.

To compare the accuracy of dedicated implementations, we show the PWC results of FPGA implementations of ViBe and PBAS algorithms. The results are shown over the CDNET-2012 [58] dataset. It is seen that EBSCam-PMM outperforms most of the methods except for ViBe. However, it will be seen shortly that the hardware complexity of ViBe is much higher compared to the proposed method (Tables 2, 4 and 6).

Table 1 PWC comparison of different methods on CDNET-2014

Category	GMM [22]	EGMM [29]	KDE [57]	ViBe [33]	PBAS [35]	EBSCam [4]	EBSCam-PMM
B. Weather	0.79	0.77	0.72	1.11	0.60	1.19	1.36
Baseline	1.53	1.32	0.55	1.08	1.88	2.05	1.88
C. Jitter	4.22	4.41	5.13	15.34	2.77	3.21	2.53
D. Background	1.21	1.17	1.64	11.57	1.00	1.41	0.86
I. O. Motion	5.19	5.49	10.07	8.75	5.23	5.15	5.53
L. Frame-rate	1.29	1.36	1.31	5.46	1.17	1.64	1.87
N. Videos	4.92	4.72	5.27	4.23	2.03	2.62	2.15
PTZ	14.53	16.94	32.31	35.37	6.15	10.32	5.34
Shadow	2.19	2.19	1.68	1.84	2.00	2.72	2.43
Thermal	4.26	4.30	1.67	2.49	4.73	5.05	5.58
Turbulence	1.27	1.24	1.51	2.12	0.22	2.11	0.74
Overall	3.77	3.99	5.14	8.13	2.53	3.41	2.75

Table 2 PWC comparison of FPGA implementations on CDNET-2012

Category	GMM [50]	ViBe [54]	PBAS [55]	EBSCam [4]	EBSCam- PMM
Baseline	1.88	N.A.	N.A	2.05	1.88
C. Jitter	5.54	N.A.	N.A	3.21	2.53
D. Background	2.19	N.A.	N.A	1.41	0.86
I. O. Motion	6.40	N.A.	N.A	5.15	5.53
Shadow	2.74	N.A.	N.A	2.72	2.43
Thermal	4.58	N.A.	N.A	5.05	5.58
Overall	3.83	1.7	3.43	3.27	3.15

The actual foreground masks generated by different algorithms for a variety of scenes are shown in Table 3. False negatives are observed when the foreground and background intensities are very similar. However, it is seen that generally EBSCam-PMM shows lower number of false positives compared to other methods. It is also seen that PBAS cannot distinguish foreground from the background as seen with the *dynamic background* sequence.

7.2 FPGA Implementation of EBSCam-PMM

To analyze the area utilization, power consumption, and speed of EBSCam-PMM for embedded applications such as smart cameras, in this subsection we discuss the FPGA implementation. We also compare the FPGA implementation of the proposed method with other implementations of BS schemes.

Table 3 Foreground masks of CDNET-2014 using EBSCam-PMM

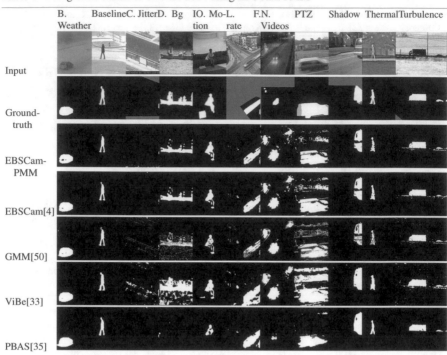

	B. Weather	BaselineC. JitterD. Bg	IO. Mo-L. tion	F.N. rate	PTZ	Shadow	ThermalTurbulence
Input				Videos			
Ground-truth							
EBSCam-PMM							
EBSCam[4]							
GMM[50]							
ViBe[33]							
PBAS[35]							

Table 4 Memory bandwidth comparison

Method	Bits per pixel	Bandwidth for 20 HD fps (Gbits/s)
EBSCam-PMM	90	7.26
GMM [50]	99	7.96
GMM [51]	108	8.65
GMM [59]	132	≥2.63 compressed
		10.51 uncompressed
ViBe [54]	160	12.67
PBAS [55]	2048	78.09

For synthesis of the circuit, we used XST. ISE was used for place and route on the FPGA. To verify the functionality of the RTL, Xilinx ISE simulation (ISIM) was used. Power estimates were obtained using XPower analyzer.

Due to increasing video resolutions and frame rates, the number of pixels that need to be processed in digital videos is continuously increasing. Thus, the memory bandwidth becomes an important feature of hardware design. Algorithms which require lower memory bandwidths are more suitable to hardware implementations. The number of bits maintained per pixel of the frame in EBSCam-PMM is 90. It is seen in Table 4 that the memory bandwidth of EBSCam-PMM is lower compared to GMM. Compared to ViBe and PBAS, the memory bandwidth of EBSCam-PMM is very small, showing the utility of the proposed method compared to these implementations. In [59], authors have used compression to reduce the memory bandwidth of GMM-based BS. However, compression adds to the circuitry, resulting in increased power consumption, area, and delay.

The implementation results of EBSCam-PMM engine on an FPGA are shown in Table 5. The background model is stored in the internal SRAM of FPGA. From these results, it is seen that EBSCam-PMM requires low area, has high speed, and consumes low power. As an example, for SVGA sequences EBSCam-PMM consumes 1.192 W at 30 fps. Also, a frame rate of approximately 96 can be achieved with EBSCam-PMM at SVGA resolution.

To compare the performance of the proposed method against previous art, we show the implementation results of our method and other methods in Table 6. It is seen that EBSCam-PMM requires significantly lower power compared to GMM. This comes as no surprise as EBSCam-PMM uses parameters that do not fluctuate rapidly, resulting in lower switching activity and, thus, lower power consumption. The logic resources of the FPGA used by EBSCam-PMM are also lower and EBSCam-PMM can achieve higher speeds compared to GMM. The logic requirement and speed of EBSCam-PMM are significantly better than ViBe and PBAS. In fact, the power consumption of EBSCam-PMM is almost negligible compared to that of PBAS. Note that EBSCam-PMM only requires slightly more logic resources compared to EBSCam, with almost similar speed and power consumption.

Recently, [61] proposed a method to reduce the memory bandwidth of GMM. From Table 7, it is seen that their method achieves a lower memory bandwidth compared to EBSCam-PMM. However, it is seen that the speed and power consumption of our implementation are much lower compared to [61]. Also, note that their method can at best achieve the accuracy of GMM, whereas it is seen earlier that EBSCam-PMM outperforms GMM in BS accuracy.

8 Conclusion

This chapter presented a new background subtraction scheme based on Poisson mixture models called EBSCam-PMM. Since shot noise is the most dominant form of noise in natural images, it is natural to use a Poisson mixture model to model the background intensity. A sequential approach to estimating the parameters of the Poisson mixture model is also presented. The estimated parameters are more robust against noise in the samples. Resultantly, the proposed method shows superior performance compare to numerous common algorithms. Furthermore, an FPGA implementation of the proposed method is also presented. EBSCam-PMM requires low memory bandwidth and battery power while providing very high frame rates at high resolutions. These features of EBSCam-PMM make it a suitable candidate for smart cameras.

Table 5 EBSCam-PMM engine on FPGA

Resolution	FPGA	LUT	FF	Slice	DSP Slice	BRAM	Frequency (MHz)	Max. fps Max. fps	Power at 30 fps (W)	Dynamic Power at 30 fps (W)
SVGA	Virtex7 xc7vx690t	837/433200	106/866400	466/108300	0/2880	1350/1470	46.34	96	1.192	0.849
VGA	Virtex7 xc7vx485t	733/303600	106/607200	439/75900	0/2800	901/1030	54.32	176	0.461	0.237
QVGA	Artix7 xc7ac200t	715/134600	105/269200	395/33560	0/740	225/365	53.48	696	0.107	0.029

Table 6 EBSCam-PMM BS circuit and comparison with previous art

FPGA	Method	Circuit	LUT	FF	Slice	DSP Slice	BRAM	Frequency (MHz)	Energy per pixel (nJ)
Virtex6 xc6vlx75t	EBSCam-PMM	Prop.	451/46560	136/93120	209/11640	0/288	1/156	270.1	0.17
	EBSCam	Ref. [4]	446/46560	120/93120	206/11640	0/288	1/156	271.3	0.14
	GMM	Ref. [50]	788/46560	363/93120	349/11640	3/288	0/156	189.3	0.72
Virtex5 xc5vlx50	EBSCam-PMM	Prop.	537/28800	119/28800	257/7200	0/48	1/48	314	0.19
	EBSCam	Ref. [4]	531/28800	111/28800	253/7200	0/48	1/48	320	0.18
	GMM	Ref. [50]	724/28800	223/28800	323/7200	3/48	3/48	130.9	1.02
		Ref. [52]	1572/28800	0/28800	N. A.	N. A.	0/48	47	5.87
		Ref. [60]	1066/28800	0/28800	346/7200	10/48	0/48	50.5	4.60
Virtex7 xc7vlx240t	EBSCam-PMM	Prop.	448/150720	128/301440	218/37680	0/768	1/832	301	0.15
	EBSCam	Ref. [4]	445/150720	120/301440	213/37680	0/768	1/832	306	0.13
	ViBe	Ref. [54]	9278/150720	12571/301440	N. A.	13/768	172/832	140	N. A.
Virtex7 xc7vlx240t	EBSCam-PMM	Prop.	532/303600	121/607200	258/75900	0/2800	1/1030	281	0.11
	EBSCam	Ref. [4]	526/303600	120/607200	254/75900	0/2800	1/1030	284	0.09
	PBAS	Ref. [55]	36060/303600	39108/607200	14903/75900	12/2800	248/1030	116	206.3

Table 7 Comparison of CoSCS-MoG and EBSCam-PMM

Method	Design	Delay/frame (ms)	Bandwidth (Gbits/s) at 20 HD fps	Energy per frame (mJ)
CoSCS-MoG	Ref. [61]	65.3	1.68	125.96
EBSCam-PMM	Prop.	1.82×10^{-5}	7.26	3.4

References

1. McIvor AM (2000) Background subtraction techniques. Proc Image Vision Comput 4:3099–3104
2. Khan MUK, Kyung CM, Yahya KM (2013) Optimized learning rate for energy waste minimization in a background subtraction based surveillance system. In: IEEE int. symp. on circuits and sys. (ISCAS). IEEE, Piscataway, pp 2355–2360
3. Khan MUK, Khan A, Kyung C-M (2014) Dual frame rate motion detection for memory-and energy-constrained surveillance systems. In: IEEE int. conf. on adv. video and signal based surveillance (AVSS). IEEE, Piscataway, pp 81–86
4. Khan MUK, Khan A, Kyung C-M (2017) Ebscam: background subtraction for ubiquitous computing. IEEE Trans Very Large Scale Integra VLSI Syst 25(1):35–47
5. Piccardi M (2004) Background subtraction techniques: a review. In: 2004 IEEE international conference on systems, man and cybernetics, vol 4. IEEE, Piscataway, pp 3099–3104
6. Kristensen F, Hedberg H, Jiang H, Nilsson P, Öwall V (2008) An embedded real-time surveillance system: implementation and evaluation. J Signal Process Syst 52(1):75–94
7. Radke RJ, Andra S, Al-Kofahi O, Roysam B (2005) Image change detection algorithms: a systematic survey. IEEE Trans Image Process 14(3):294–307
8. Seki M, Wada T, Fujiwara H, Sumi K (2003) Background subtraction based on cooccurrence of image variations. In: Proceedings 2003 IEEE computer society conference on computer vision and pattern recognition, 2003, vol 2. IEEE, Piscataway, pp II–II
9. Power PW, Schoonees JA (2002) Understanding background mixture models for foreground segmentation. In: Proceedings image and vision computing, New Zealand, vol 2002, pp 10–11
10. Tsai D-M, Lai S-C (2009) Independent component analysis-based background subtraction for indoor surveillance. IEEE Trans Image Process 18(1):158–167
11. Lin H-H, Liu T-L, Chuang J-H (2009) Learning a scene background model via classification. IEEE Trans Signal Process 57(5):1641–1654
12. Sivabalakrishnan M, Manjula D (2009) An efficient foreground detection algorithm for visual surveillance system. Int J Comput Sci Netw Secur 9(5):221–227
13. Jacques J, Jung CR, Musse SR (2006) A background subtraction model adapted to illumination changes. In: Proc. IEEE int. conf. on image processing, pp 1817–1820
14. Davis JW, Sharma V (2004) Robust background-subtraction for person detection in thermal imagery. In: IEEE int. workshop on object tracking and classification beyond the visible spectrum
15. Wren CR, Azarbayejani A, Darrell T, Pentland AP (1997) Pfinder: real-time tracking of the human body. IEEE Trans Pattern Anal Mach Intell 19(7):780–785
16. Manzanera A, Richefeu JC (2007) A new motion detection algorithm based on $\sigma - \delta$ background estimation. Pattern Recogn Lett 28(3):320–328
17. Manzanera A (2007) σ-δ background subtraction and the Zipf law. In: Progress in pattern recognition, image analysis and applications, pp 42–51
18. Lacassagne L, Manzanera A, Denoulet J, Mérigot A (2009) High performance motion detection: some trends toward new embedded architectures for vision systems. J Real-Time Image Process 4(2):127–146
19. Lacassagne L, Manzanera A (2009) Motion detection: fast and robust algorithms for embedded systems. In: Proc. IEEE int. conf. on image processing

20. Elgammal A, Harwood D, Davis L (2000) Non-parametric model for background subtraction. In: Computer VisionECCV 2000, pp 751–767
21. Tavakkoli A, Nicolescu M, Bebis G, Nicolescu M (2009) Non-parametric statistical background modeling for efficient foreground region detection. Mach Vis Appl 20(6):395–409
22. Stauffer C, Grimson WEL (1999) Adaptive background mixture models for real-time tracking. In: IEEE conf. on comp. vis. and pattern recog., vol 2. IEEE, Piscataway
23. White B, Shah M (2007) Automatically tuning background subtraction parameters using particle swarm optimization. In: 2007 IEEE international conference on multimedia and expo. IEEE, Piscataway, pp 1826–1829
24. Atrey PK, Kumar V, Kumar A, Kankanhalli MS (2006) Experiential sampling based foreground/background segmentation for video surveillance. In: 2006 IEEE international conference on multimedia and expo. IEEE, Piscataway, pp 1809–1812
25. Park J, Tabb A, Kak AC (2006) Hierarchical data structure for real-time background subtraction. In: 2006 IEEE international conference on image processing. IEEE, Piscataway, 1849–1852
26. Porikli F, Tuzel O (2003) Human body tracking by adaptive background models and mean-shift analysis. In: IEEE international workshop on performance evaluation of tracking and surveillance, pp 1–9
27. Yang S-Y, Hsu C-T (2006) Background modeling from GMM likelihood combined with spatial and color coherency. In: 2006 IEEE international conference on image processing. IEEE, Piscataway, pp 2801–2804
28. Lee D-S (2005) Effective gaussian mixture learning for video background subtraction. IEEE Trans Pattern Anal Mach Intell 27(5):827–832
29. Zivkovic Z (2004) Improved adaptive gaussian mixture model for background subtraction. In: Proc. IEEE int. conf. on pattern recognition, vol 2, pp 28–31
30. Wang R, Bunyak F, Seetharaman G, Palaniappan K (2014) Static and moving object detection using flux tensor with split gaussian models. In: Proc. IEEE conf. on computer vision and pattern recognition Wkshps., pp 414–418
31. Kim K, Chalidabhongse TH, Harwood D, Davis L (2005) Real-time foreground–background segmentation using codebook model. Real-Time Imaging 11(3):172–185
32. Sigari MH, Fathy M (2008) Real-time background modeling/subtraction using two-layer codebook model. In: Proceedings of the international multiconference of engineers and computer scientists, vol 1
33. Barnich O, Van Droogenbroeck M (2011) Vibe: a universal background subtraction algorithm for video sequences. IEEE Trans Image Process 20(6):1709–1724
34. Haque M, Murshed M (2013) Perception-inspired background subtraction. IEEE Trans Circuits Syst Video Technol 23(12):2127–2140
35. Hofmann M, Tiefenbacher P, Rigoll G (2012) Background segmentation with feedback: the pixel-based adaptive segmenter. In: Proc. IEEE conf. on computer vision and pattern recognition wkshps. IEEE, Piscataway, pp 38–43
36. Wang H, Suter D (2006) Background subtraction based on a robust consensus method. In: 18th international conference on pattern recognition, 2006. ICPR 2006, vol 1. IEEE, Piscataway, pp 223–226
37. Guillot C, Taron M, Sayd P, Pham QC, Tilmant C, Lavest J-M (2010) Background subtraction adapted to PTZ cameras by keypoint density estimation. In: Proceedings of the British machine vision conference, p 34-1
38. Tavakoli HR, Rahtu E, Heikkilä J (2012) Temporal saliency for fast motion detection. In: Asian conference on computer vision. Springer, Berlin, pp 321–326
39. Hamid R, Sarma AD, DeCoste D, Sundaresan N (2015) Fast approximate matching of videos from hand-held cameras for robust background subtraction. In: 2015 IEEE winter conference on applications of computer vision (WACV). IEEE, Piscataway, pp 294–301
40. Foresti GL (1998) A real-time system for video surveillance of unattended outdoor environments. IEEE Trans Circuits Syst Video Technol 8(6):697–704

41. Kalpana Chowdary M, Suparshya Babu S, Susrutha Babu S, Khan H (2013) FPGA implementation of moving object detection in frames by using background subtraction algorithm. In: 2013 International conference on communications and signal processing (ICCSP). IEEE, Piscataway, pp 1032–1036
42. Lee Y, Kim Y-G, Lee C-H, Kyung C-M (2014) Memory-efficient background subtraction for battery-operated surveillance system. In: The 18th IEEE international symposium on consumer electronics (ISCE 2014). IEEE, Piscataway, pp 1–2
43. Chan W-K, Chien S-Y (2007) Real-time memory-efficient video object segmentation in dynamic background with multi-background registration technique. In: IEEE 9th workshop on multimedia signal processing, 2007. MMSP 2007. IEEE, Piscataway, pp 219–222
44. Gujrathi P, Priya RA, Malathi P (2014) Detecting moving object using background subtraction algorithm in FPGA. In: 2014 Fourth international conference on advances in computing and communications (ICACC). IEEE, Piscataway, pp 117–120
45. Menezes GGS, Silva-Filho AG (2010) Motion detection of vehicles based on FPGA. In: Programmable logic conference (SPL), 2010 VI Southern. IEEE, Piscataway, pp 151–154
46. Cherian S, Senthil Singh C, Manikandan M (2014) Implementation of real time moving object detection using background subtraction in FPGA. In: 2014 International conference on communications and signal processing (ICCSP). IEEE, Piscataway, pp 867–871
47. Hiraiwa J, Vargas E, Toral S (2010) An FPGA based embedded vision system for real-time motion segmentation. In: Proceedings of 17th international conference on systems, signals and image processing, Brazil
48. Abutaleb MM, Hamdy A, Abuelwafa ME, Saad EM (2009) FPGA-based object-extraction based on multimodal σ-δ background estimation. In: 2nd International conference on computer, control and communication, 2009. IC4 2009. IEEE, Piscataway, pp 1–7
49. Tabkhi H, Sabbagh M, Schirner G (2014) A power-efficient FPGA-based mixture-of-Gaussian (MoG) background subtraction for full-HD resolution. In: Proc. annual int. symp. on field-programmable custom computing machines, pp 241–241
50. Genovese M, Napoli E (2014) ASIC and FPGA implementation of the gaussian mixture model algorithm for real-time segmentation of high definition video. IEEE Trans Very Large Scale Integr VLSI Syst 22(3):537–547
51. Genovese M, Napoli E (2013) FPGA-based architecture for real time segmentation and denoising of HD video. J Real-Time Image Process 8(4):389–401
52. Genovese M, Napoli E, Petra N (2010) OpenCV compatible real time processor for background foreground identification. In: Proc. IEEE int. conf. on microelectronics, pp 467–470
53. Rodriguez-Gomez R, Fernandez-Sanchez EJ, Diaz J, Ros E (2015) Codebook hardware implementation on FPGA for background subtraction. J Real-Time Image Process 10(1):43–57
54. Kryjak T, Gorgon M (2013) Real-time implementation of the ViBe foreground object segmentation algorithm. In: Proc. IEEE federated conf. on computer science and information sys., pp 591–596
55. Kryjak T, Komorkiewicz M, Gorgon M (2013) Hardware implementation of the PBAS foreground detection method in FPGA. In: Proc. IEEE int. conf. mixed design of integrated circuits and sys., pp 479–484
56. Wang Y, Jodoin P-M, Porikli F, Konrad J, Benezeth Y, Ishwar P (2014) CDnet 2014: an expanded change detection benchmark dataset. In: Proc. IEEE conf. on computer vision and pattern recognition wkshps., pp 387–394
57. Elgammal A, Duraiswami R, Harwood D, Davis LS (2002) Background and foreground modeling using nonparametric kernel density estimation for visual surveillance. Proc IEEE 90(7):1151–1163
58. Goyette N, Jodoin P-M, Porikli F, Konrad J, Ishwar P (2012) Changedetection. net: a new change detection benchmark dataset. In: Proc. IEEE conf. on computer vision and pattern recognition wkshps., pp 1–8

59. Jiang H, Ardö H, Öwall V (2009) A hardware architecture for real-time video segmentation utilizing memory reduction techniques. IEEE Trans Circuits Syst Video Technol 19(2):226–236
60. Ratnayake K, Amer A (2014) Embedded architecture for noise-adaptive video object detection using parameter-compressed background modeling. J Real-Time Image Process, pp 1–18
61. Shen Y, Hu W, Yang M, Liu J, Wei B, Lucey S, Chou CT (2016) Real-time and robust compressive background subtraction for embedded camera networks. IEEE Trans Mobile Comput 15(2):406–418

Key-Frame SLAM Based on Motion Estimation and Stochastic Filtering Using Stereo Vision

Jungho Kim, Youngbae Hwang, and In So Kweon

1 Introduction

Simultaneous Localization and Mapping (SLAM) is the problem of building up a global map of an unknown environment while simultaneously estimating the current pose of a robot (or a sensor) with given sensor measurements. For decades, there has been intensive research on *visual SLAM* that primarily uses visual sensors such as cameras for this task. This problem has attracted immense attention in the robotics and computer vision communities. Technically, visual SLAM can be defined as the problem of estimating the posterior distribution of the current state vector represented in the probabilistic form as

$$p(x_t | z_{1:t}, u_{1:t}), \tag{1}$$

where $z_{1:t}$ and $u_{1:t}$ are given visual measurements and control inputs obtained up to time t, respectively, and x_t is the current state vector constructed by the camera pose and 3D locations of observed landmarks.

To recursively estimate the posterior distribution of the current state in Eq. (1) over time, we decompose it into the measurement model, the process model, and the posterior distribution of the previous state using the Bayes' theorem as follows:

J. Kim
Image Processing Research Center, KETI, Seongnam-si, Gyeonggi-do, South Korea
e-mail: jhkim77@keti.re.kr

Y. Hwang (✉)
Electronics Engineering, Chungbuk National University, Cheongju, Chungbuk, South Korea
e-mail: ybhwang@cbnu.ac.kr

I. S. Kweon
Electrical Engineering, KAIST, Daejeon, Chungcheongnam-do, South Korea
e-mail: iskweon@kaist.ac.kr

© Springer Nature Switzerland AG 2020
Y. Liu et al. (eds.), *Smart Sensors and Systems*,
https://doi.org/10.1007/978-3-030-42234-9_2

$$p(x_t|z_{1:t}, u_{1:t}) = \eta p(z_t|x_t) \int_{x_{t-1}} p(x_t|x_{t-1}, u_t)p(x_{t-1}|z_{1:t-1}, u_{1:t-1})dx_{t-1}. \quad (2)$$

Here, η is a normalizing constant.

In Eq. (2), we have two probabilistic models: the process model $p(x_t|x_{t-1}, u_t)$ and the measurement model $p(z_t|x_t)$ [1, 2]. In the process model, the current state x_t is governed by the probabilistic function of the previous state x_{t-1}, the control input u_t, and the additive process noise w_t as:

$$x_t = f(x_{t-1}, u_t, w_t). \quad (3)$$

The process noise w_t is the statistical description of the errors in the process model.

The measurements z_t are also governed by the probabilistic function of the current state and the additive measurement noise v_t as

$$z_t = h(x_t, v_t). \quad (4)$$

Here, two random variables, w_t and v_t, representing the process noise and the measurement noise, respectively, are assumed to be independent as shown in Fig. 1 [2, 3]. To ensure the independence between the process noise and the measurement noise, a constant-velocity model independent of visual data has been employed for the control input u_t involved in the process model in many visual SLAM approaches [4–6]. However, if a camera undergoes sudden motion changes, a constant-velocity model is not valid any more and yields poor results. On the other hand, some other sensors such as a wheel encoder [7] and an inertial measurement unit (IMU) [8] are used to obtain control inputs. However, wheel encoders cannot be used to obtain the 6-degree-of-freedom (DOF) poses and IMUs suffer from an error accumulation problem. In this sense, global positioning system (GPS) sensors are

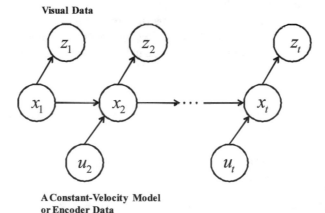

Fig. 1 Bayesian network describing the relation between process and measurement models

Fig. 2 The GPS coordinates obtained near the buildings show its weakness for jamming problems (Alignment with the Google map was done with a program provided by a GPS manufacturer)

powerful for localization, because they can avoid an error accumulation problem. However, their performance is strongly affected by environment conditions: for example, if a GPS is located near the buildings, its output is not reliable as shown in Fig. 2. Moreover, a GPS cannot be used for indoor environments.

In contrast, motion estimation approaches can be adopted to reliably estimate the 6-DOF pose under erratic camera motion. However, if we use motion estimation methods for the process model, then the control input u_t directly depends on measurements z_t and the independence assumption of the process noise and the measurement noise is no longer valid.

To resolve this problem, we divide visual measurements (i.e., image features) into two categories—*common features* consistently observed in the consecutive key-frame images and *new features* newly detected in the current key-frame image. Two groups of features are used for process and measurement models, separately. Here, we focus on the uncertainty or noise of feature detection and association for vision sensors, even though visual measurements for visual SLAM are also corrupted by various factors such as system calibration errors, quantization errors, and image blur due to motion, etc.

In Eq. (2), we first predict the distribution of the current state from the process model, and then correct it by using the measurement model. As a camera moves, the information about its new pose is predicted. Here, the problem is that the accuracy naturally decreases and the uncertainties equivalently increase at the same time. However, when some of the landmarks are re-observed in the images, their observations are used for updating the camera pose estimation and improving the estimation of observed landmark locations. Therefore, those observations decrease the uncertainties of the camera pose and landmark locations.

In this sense, we formulate a key-frame-based Bayesian filtering framework for visual SLAM. In the proposed framework, the camera path is composed of key-frame poses that are effectively estimated by observations at non-key-frame locations. Thus, our key-frame SLAM approach reduces the number of predictions, which increase the uncertainties, by reducing the number of camera poses to be estimated in the path. In addition, we update the distribution of the camera path by using many observations obtained at the non- key-frame locations to decrease the uncertainties further.

For efficient estimation of 3D landmarks, we employ a FastSLAM [9]-type approach in which each 3D landmark is individually estimated with the given camera path.

Effectively approximating the distribution of the 6-DOF camera pose is a challenging task in the particle filter-based localization or FastSLAM because a particle filter, which constructs a sample-based representation of the entire distribution, requires many particles of the variable of interest even when the dimension of state increases. Consequently, it becomes infeasible to manage enough particles to represent the posterior distribution of the state.

To effectively approximate the distribution of the relative motion by a limited number of particles, we directly use some hypotheses coming from motion estimation.

In summary, our contribution is twofold:

- We achieve the independence between the process noise and the measurement noise, which is a critical factor when using vision-based motion estimation approaches for the process model.
- We present a novel key-frame SLAM approach to reduce the number of camera poses to be estimated in the path. In addition, we propose key-frame-based Bayesian filtering to effectively update the posterior distribution of the camera path by using many observations obtained at non- key-frame locations.

The remainder of this paper is organized as follows: Section 2 introduces the related works. Section 3 describes the details of our visual odometry for motion estimation. Section 4 depicts our key-frame-based Bayesian filtering to compensate for errors involved in sensor measurements and motion estimation. Section 5 shows various experimental results and the evaluation of the proposed method. We will then conclude the paper in Sect. 6.

2   Related Works

To solve the SLAM problem, many methods based on the recursive estimation of the posterior distribution have been introduced. Davison [4, 10] successfully performed monocular SLAM by employing an extended Kalman filter (EKF) and adopting an initialization process to determine the depth of the 3D landmarks by using the particle filter-type approach. In [9], the authors proposed FastSLAM which factorizes the full posterior into a product of conditional landmark distributions and a distribution over camera paths. This algorithm is an instance of the Rao-Blackwellized particle filter [11, 12] that results in a substantial computational gain as one only has to sample some of variables and apply closed-form filtering such as Kalman filtering to lower dimensional sub-networks for the rest of variables. Kim et al. [13] proposed a SLAM algorithm based on unscented transformation called unscented FastSLAM (UFastSLAM) which overcomes drawbacks of the Rao-Blackwellized particle filter caused by non-linear relations. Eade and Drummond [14] utilized the FastSLAM-type particle filter in single-camera SLAM to manage a greater number of landmarks because the computational requirements of EKF-based SLAM approaches rapidly grow with the number of landmarks. In [5], the authors described a visual SLAM algorithm that is robust to erratic camera motion and visual occlusion by using efficient scale prediction and exaemplar-based feature representations in conjunction with the use of an unscented Kalman filter (UFK) [15]. Recently, Eade et al. [6] proposed a monocular SLAM system in which map inconsistencies can be prevented by coalescing observations into independent local coordinate frames, building a graph of the local frames, and optimizing the resulting graph. This approach is based on hierarchical optimization in that local states are updated using local bundle adjustment and multiple observations sharing between local coordinates are used for updating nodes (local coordinate frames) by graph optimization. For the similar problem with Eade's work, Paz et al. [16] presented a 6-DOF visual SLAM system based on conditionally independent local maps and the strategy of updating the global map by using observations sharing between local maps instead of directly estimating the global map. Here, it is worth noting that in most 3D visual SLAM approaches, a constant-velocity model is employed to achieve the independence between the process and measurement noise. However, in the constant velocity model, when cameras undergo sudden motion, these SLAM approaches are highly prone to fail, resulting in inconsistencies in the global map. In [17], the authors combined the particle filter-based localization with the UKF-based SLAM to cope with erratic camera motion while maintaining a small number of landmarks.

On the other hand, in the vision community, structure-from-motion (SFM) approaches [18] have been studied independent of SLAM to estimate camera trajectories by using a sequence of images only. For example, Nister et al. introduced "'visual odometry'" that estimates the relative movements of the stereo head in the Euclidean space [19]. Recently, Zhu et al. [20] developed a helmet-based visual odometry system that consists of two pairs of stereo cameras mounted on a helmet; one pair faces forward while the other faces backward. By utilizing the multi-stereo fusion algorithm, they improved the overall accuracy in pose estimation. Here, we should note that, in many previous studies, optimization techniques such as bundle adjustment [21, 22] have been adopted to avoid inconsistencies in the global map. However, it is not feasible to perform conventional bundle adjustment in on-line approaches because the computational cost rapidly grows with the number of 3D landmarks and their observations (the image coordinates over the sequence). Recently, Jeong et al. [23] proposed a fast method for bundle adjustment by using block-based preconditioned conjugate gradient and embedded point iterations. In addition, to achieve the computational efficiency and the consistency of the global map, local bundle adjustment [24, 25] and hierarchical bundle adjustment [26] techniques have been studied. In [27], FrameSLAM was presented using non-linear least-squares estimation for local registration and loop closure that result in accurate maps.

3   The Visual Odometry System

Our visual odometry system consists of a few sub-components including feature matching, motion estimation, and key-frame selection.

3.1   Feature Extraction and Stereo Matching

For each stereo pair, we first extract corner points in the left image and then apply the 1D KLT feature tracker [28] to stereo images to obtain correspondences. Because the KLT tracker gives the sub pixel locations of matched points, we can reconstruct more accurate structures. We assume that a point (x, y) in the left image I corresponds to a point $(x - d_{min} - d_x, y)$ in the right image J, and we linearize $J(x - d_{min} - d_x, y)^1$ by Taylor expansion as

$$I(x, y) \simeq J(x - d_{min}, y) - g_x d_x., \tag{5}$$

Here, d_{min} is a minimum disparity which makes d_x smaller to satisfy the first order approximation of Taylor expansion, and g_x is an intensity gradient along the horizontal axis. We determine the disparity d_x that minimizes the following dissimilarity measure:

$$\epsilon = \int \int_A [h(x, y) - g_x d_x]^2 w \, dA, \tag{6}$$

where A denotes a local mask, w is a weight within the local mask, and $h(x, y) = J(x - d_{min}, y) - I(x, y)$. To find the disparity d_x, we set the derivative of Eq. (6) w.r.t. d_x to zero. Finally, d_x is computed by

$$d_x = \frac{\int \int_A h(x, y) g_x w \, dA}{\int \int_A g_x^2 w \, dA}. \tag{7}$$

---

[1]$I(x, y)$ represents an intensity value of a point (x, y) in the left image I and $J(x - d_{min} - d_x, y)$ represents an intensity value of a point $(x - d_{min} - d_x, y)$ in the right image J.

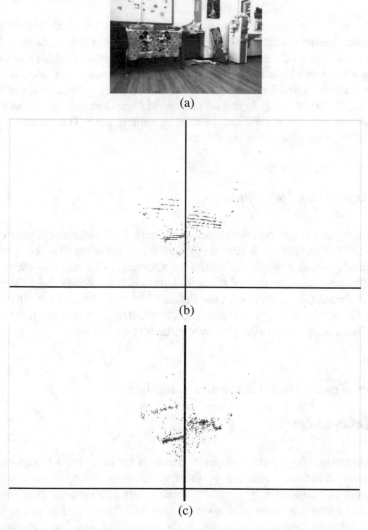

Fig. 3 Comparison of 3D reconstruction. (**a**) Left image from stereo. (**b**) The top-down view of 3D reconstruction by NCC stereo matching. (**c**) The top-down view of 3D reconstruction by 1D KLT stereo matching

Figure 3 shows the top views of 3D reconstruction computed from NCC and 1D-KLT stereo matching methods, respectively.

The 3D coordinates of the matched corner points are used for map building and for motion estimation.

3.2   Motion Estimation

Nister et al. [19] introduced a method called "'visual odometry'" for the real-time estimation of the movement of a stereo head or a single camera. In this approach, the 3-point algorithm [29] is employed: the images of three known world points are used to obtain up to four possible camera poses (one solution can be automatically obtained by using more than three point). Here, we additionally employ the RANSAC [30] where a set of 3 world points and their image coordinates are randomly selected to compute the relative camera pose. The estimated pose is evaluated by using other correspondences.

3.3   Key-Frame Designation

The number of tracked corners between an incoming image and a previous key image can be a measure to determine the key-frame locations. In [31], the authors automatically select key- frames suited for structure and motion recovery on the basis of the number of the tracked features. Similarly, if the number of points tracked by the KLT tracker is smaller than a pre-defined threshold, we designate an incoming image as a key-frame image and estimate the relative pose w.r.t the previous key-frame. This strategy can partially prevent error accumulation.

4   Key-Frame-Based Bayesian Filtering

4.1   Independency

When we use existing motion estimation methods for the process model, the noise in motion estimation is affected by the measurement noise, which renders the assumption on independency invalid. To solve this problem, we first divide the observations into two categories when a new key-frame images is determined: common features consistently observed in the consecutive key-frame images z^c and newly detected features in the current key-frame image z^d ($= z - z^c$), as shown in Fig. 4. All the features are generally independent of each other because they are individually extracted and tracked. z^c are used to evaluate the posterior distribution of the state x_t, and z^d are used to estimate the relative pose, u_t, as the control input involved in the process model.

We can re-define the process model and the measurement model from two sets of measurements as shown in Eq. (8). This means that the process noise w_t is only dependent on the measurements z_t^d and the measurement noise v_t is determined by the measurements z_t^c.

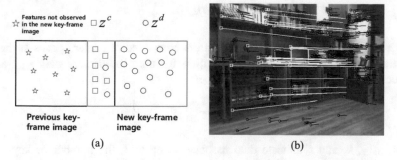

Fig. 4 Achieving independence between process noise and measurement noise by dividing observations. (**a**) Consecutive key-frame images. (**b**) Black circles and white rectangles represent features belonging to z_t^d and z_t^c, respectively

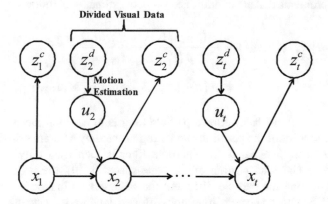

Fig. 5 Bayesian network that describes the independence between process and measurement noise in the case of divided observations. There is no closed path within the directed graph

$$x_t = f\left(x_{t-1}, z_t^d, w_t\right)$$
$$z_t^c = h\left(x_t, v_t\right) \tag{8}$$

We then achieve the independence between the process noise and measurement noise by simply dividing the observations instead of using another sensor as shown in Fig. 5. Thus the posterior distribution defined by $u_{1:t}$ and $z_{1:t}$ in Eq. (2) can be formulated with two sets of image features, $z_{1:t}^c$ and $z_{1:t}^d$, as

$$p(x_t|z_{1:t}^c, z_{1:t}^d) = p(z_t^c|x_t) \int_{x_{t-1}} p(x_t|x_{t-1}, z_t^d)p(x_{t-1}|z_{1:t-1}^c, z_{1:t-1}^d)dx_{t-1}. \tag{9}$$

4.2 Proposed Target Distribution

In the proposed method, the camera path is composed of camera poses at the key frames. In this case, our target distribution for both the camera path $n_{1:k}$ and a collection of landmarks M is expressed as

$$p\left(n_{1:k}, M | z_{1:t}^c, z_{1:t}^d, d_{1:t}\right), \tag{10}$$

where $d_{1:t}$, k, and t indicate data association up to t, the number of key-frame locations, and the number of all the frames, respectively.

By employing a Rao-Blackwellized particle filtering technique [11, 12], a SLAM problem is decomposed into a localization problem and a collection of landmark estimation problems that are conditioned on path estimates as follows:

$$p\left(n_{1:k}, M | z_{1:t}^c, z_{1:t}^d, d_{1:t}\right) = \frac{p\left(n_{1:k}, M, z_{1:t}^c, z_{1:t}^d, d_{1:t}\right)}{p\left(z_{1:t}^c, z_{1:t}^d, d_{1:t}\right)}$$
$$= p\left(n_{1:k} | z_{1:t}^c, z_{1:t}^d, d_{1:t}\right) p\left(M | n_{1:k}, z_{1:t}^c, z_{1:t}^d, d_{1:t}\right). \tag{11}$$

The idea of a Rao-Blackwellized particle filter is to partition the state vector so that one component of the partition can be represented by a parametric distribution that is analytically estimated. The particle filter is then used only for the non-linear and non-Gaussian portion of the state-space. In [9], the path estimator is implemented using the particle filter and the landmark estimator is implemented using an EKF with a separate filter for different landmarks, because all the 3D landmarks are independent of each other with a given path as

$$p\left(M | n_{1:k}, z_{1:t}^c, z_{1:t}^d, d_{1:t}\right) = \prod_{l=1}^{L} p\left(m_l | n_{1:k}, z_{1:t}^c, z_{1:t}^d, d_{1:t}\right), \tag{12}$$

where m_l represents each landmark in M.

4.3 Key-Frame-Based Path Estimation

We estimate the distribution of the camera path, $p\left(n_{1:k} | z_{1:t}, d_{1:t}\right)$, that consists of a sequence of camera poses at only the key-frame locations instead of all the frames, as shown in Fig. 6. When designating an incoming image as a key-frame image, we elongate the path, $n_{1:k+1}$, by adding the relative pose, u_t, with respect to the last key-frame pose, n_k, to the previous path, $n_{1:k}$.

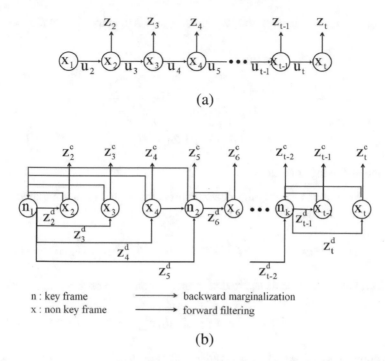

n : key frame   ⟶   backward marginalization
x : non key frame   ⟶   forward filtering

(b)

Fig. 6 State transition. (**a**) State transition for sequential SLAM—it recursively estimates the current state by using incoming observations. Where x_t is the state, u_t is an encoder input, and z_t is a sensor measurement at time t. (**b**) State transition for proposed (key-frame-based) SLAM—it recursively estimates the state consisting of key frames using incoming observations. where n_k is the state corresponding to the kth key frame, z_t^d is a set of measurements for the process model, z_t^c is a set of measurements for the measurement model, and x_t is the state for non- key frames

The posterior distribution of $n_{1:k}$ with the given $z_{1:t} (= z_{1:t}^c + z_{1:t}^d)$ and $d_{1:t}$ (an indicator for correspondences between the landmarks M, and observations $z_{1:t}^c$) is estimated by marginalizing out the relative pose u_t from the joint distribution as

$$p\left(n_{1:k} | z_{1:t}^c, z_{1:t}^d, d_{1:t}\right) = \int_{u_t} p\left(n_{1:k}, u_t | z_{1:t}^c, z_{1:t}^d, d_{1:t}\right) du_t. \qquad (13)$$

This means that u_t is a latent variable that is marginalized out to estimate the distribution of $n_{1:k}$ at non- key-frame locations. The joint distribution $p(n_{1:k}, u_t | z_{1:t}^c, z_{1:t}^d, d_{1:t})$ estimated up to time t is used for updating the previous camera path defined at key-frame locations to re-estimate the posterior distribution of the camera path $n_{1:k}$, and that posterior distribution is updated by using observations at the non-key-frame locations unless the new key- frame is designated. Also, the prediction of the current state is performed from the previous key-frame location, not the location of the previous frame, as shown in Fig. 6, because visual odometry that is used for the process model provides the relative pose with respect to the previous key-frame location.

Using the Bayes' theorem, the joint distribution is decomposed as follows:

$$p\left(n_{1:k}, u_t \mid z_{1:t}^c, z_{1:t}^d, d_{1:t}\right)$$

$$= \frac{p\left(n_{1:k}, u_t, z_t^c, z_t^d, d_t, z_{1:t-1}^c, z_{1:t-1}^d, d_{1:t-1}\right)}{p\left(z_{1:t}^c, z_{1:t}^d, d_{1:t}\right)} \tag{14}$$

$$= \eta p\left(z_t^c \mid n_{1:k}, u_t, d_t\right) p\left(u_t \mid z_t^d\right) p\left(n_{1:k} \mid z_{1:t-1}^c, z_{1:t-1}^d, d_{1:t-1}\right)$$

under the following conditions:,

(a) The observations z_t^c are only dependent on the global pose computed by $n_{1:k}$ and u_t and current data association d_t.

$$p\left(z_t^c \mid n_{1:k}, u_t, d_t\right) = p\left(z_t^c \mid n_{1:k}, u_t, z_t^d, d_t, z_{1:t-1}^c, z_{1:t-1}^d, d_{1:t-1}\right)$$

(b) The relative movements u_t are determined by only the observations z_t^d.

$$p\left(u_t \mid z_t^d\right) = p\left(u_t \mid n_{1:k}, z_t^d, d_t, z_{1:t-1}^c, z_{1:t-1}^d, d_{1:t-1}\right)$$

(c) Without observations z_t^c, $n_{1:k}$ is independent of d_t.

$$p\left(n_{1:k} \mid z_{1:t-1}^c, z_{1:t-1}^d, d_{1:t-1}\right) = p\left(n_{1:k} \mid z_{1:t-1}^c, z_{1:t-1}^d, d_t, d_{1:t-1}\right)$$

where $p(z_t^c \mid n_{1:k}, u_t, d_t)$ is a likelihood for the measurement model only depending on the measurements z_t^c, and $p(u_t \mid z_t^d)$ is the posterior distribution of the relative pose for the process model depending on the measurements z_t^d, as defined by Eqs. (19) and (16), respectively. $p(n_{1:k} \mid z_{1:t-1}, d_{1:t-1})$ is the previous posterior distribution up to $t - 1$.

Thus our key-frame SLAM approach can reduce the number of camera poses to be estimated in the path by generating the poses only at key-frame locations and effectively update the posterior distribution of the camera path by using many observations obtained at non- key-frame locations.

4.4 Non-parametric Representation

The posterior distribution $p(n_{1:k} \mid z_{1:t}^c, z_{1:t}^d, d_{1:t})$ is represented by a set of weighted particles, as shown below:.

$$p\left(n_{1:k} \mid z_{1:t}^c, z_{1:t}^d, d_{1:t}\right) = \sum_i p\left(n_{1:k}^i \mid z_{1:t}^c, z_{1:t}^d, d_{1:t}\right) \delta\left(n_{1:k} - n_{1:k}^i\right), \tag{15}$$

where $\delta(x)$ represents the Dirac delta function that returns 1 if x is zero, and 0 otherwise.

The posterior distribution of the relative pose is also approximated by a set of particles coming from the RANSAC, where relative poses are estimated by selecting multiple sets of minimal correspondences. (For the 3-point algorithm, three correspondences are required.) Each pair of the minimal set provides a single hypothesis on the relative pose, and its weight is computed according to the number of inliers among all correspondences. Thus, in our approach, we use multiple hypotheses that are generated in the RANSAC step. The RANSAC is an efficient technique for determining a good hypothesis, but unfortunately the hypothesis selected with the best score (the number of inliers) does not always correspond to the correct estimate. Therefore, instead of selecting an unique hypothesis, we propagate multiple reasonable hypotheses to the subsequent frames to re-estimate the posterior distribution by using more observations. We represent the posterior distribution of the relative pose (6 DOF) using the hypotheses and their weights according to the number of inliers as

$$
p\left(u_t^j \mid z_t^d\right) \propto \frac{N_{inlier}^j}{N_{total}}, \quad \sum_j p\left(u_t^j \mid z_t^d\right) = 1, \tag{16}
$$

where N_{inlier}^j is the number of inliers for u_t^j, and N_{total} is the total number of correspondences in z_t^d.

Thus, we compute the weight for each particle by marginalizing out the relative pose as

$$
p\left(n_{1:k}^i \mid z_{1:t}^c, z_{1:t}^d, d_{1:t}\right) = \sum_j p\left(n_{1:k}^i, u_t^j \mid z_{1:t}^c, z_{1:t}^d, d_{1:t}\right). \tag{17}
$$

We compute the joint probability of a camera path, $n_{1:k}^i$, and a relative pose, u_t^j, using Eq. (18) that is based on Eq. (14).

$$
\begin{aligned}
p\left(n_{1:k}^i, u_t^j \mid z_{1:t}^c, z_{1:t}^d, d_{1:t}\right) = {} & \eta p\left(z_t^c \mid n_{1:k}^i, u_t^j, d_t\right) p\left(u_t^j \mid z_t^d\right) \\
& \times p\left(n_{1:k}^i \mid z_{1:t-1}^c, z_{1:t-1}^d, d_{1:t-1}\right)
\end{aligned} \tag{18}
$$

4.5 Likelihood Estimation and Outlier Rejection

The likelihood estimation is based on the number of inliers for each particle. It is computed by examining how many scene points m_l are projected close to relevant measurements z_t^l belonging to z_t^c as

$$p\left(z_t^c | n_{1:k}^i, u_t^j, d_t\right) = \int_M p\left(z_t^c, M | n_{1:k}^i, u_t^j, d_t\right) dM$$

$$= \int_M p\left(z_t^c | M, n_{1:k}^i, u_t^j, d_t\right) p\left(M | n_{1:k}^i, u_t^j, d_t\right) dM \quad (19)$$

$$= \frac{1}{L} \sum_{l=1}^{L} d\left(z_t^l, m_l, n_{1:k}^i, u_t^j\right),$$

where

$$d\left(z_t^l, m_l, n_{1:k}^i, u_t^j\right) = \begin{cases} 1 & \text{if } \left\| z_t^l - z\left(\hat{m}_l, \left(n_{1:k}^i \oplus u_t^j\right)\right)\right\| < \sigma_l \\ 0 & \text{otherwise} \end{cases}$$

$$\sigma_l = \sqrt{e_0^2 + e_l^2}.,$$

Here, e_0 is a pre-defined observation uncertainty for the likelihood and e_l represents the uncertainty of the landmark m_l in the image space, computed by Eq. (22). $\hat{m}_l$ is the updated 3D location of the landmark as shown in Sect. 4.7. $(n_{1:k}^i \oplus u_t^j)$ indicates the global pose of the camera computed from the path $n_{1:k}^i$ and the relative pose u_t^j. $d(z_t^l, m_l, n_{1:k}^i, u_t^j)$ indicates whether the point m_l is an inlier or outlier with respect to the observation z_t^l and the camera pose $(n_{1:k}^i \oplus u_t^j)$, and $z(\hat{m}_l, (n_{1:k}^i \oplus u_t^j))$ is the projection of a scene point $\hat{m}_l$ for a particular camera pose $(n_{1:k}^i \oplus u_t^j)$. L is the number of scene points that are associated with the current measurements, as defined by d_t.

We eliminate the outliers z_t^o among z_t^l that are not supported by any particles in the computation of the likelihood values as

$$z_t^o = \left\{ z_t^l | \sum_i \sum_j d\left(z_t^l, m_l, n_{1:k}^i, u_t^j\right) = 0 \right\}. \quad (20)$$

Thus, we eliminate the outliers in z_t^d using the RANSAC when estimating the relative pose, u_t, and the outliers in z_t^c when computing the likelihood values.

4.6 Path Generation

Whenever we have a new key-frame image, we elongate the path using the previous posterior distribution of the camera path and the relative pose as follows:

$$n_{1:k+1}^{N_j \times i+j} \leftarrow \left\{ n_{1:k}^i, \left(n_{1:k}^i \oplus u_t^j \right) \right\},$$

$$p\left(n_{1:k+1}^{N_j \times i+j} | z_{1:t}, d_{1:t} \right) \propto p\left(u_t^j | z_t^d \right) p\left(n_{1:k}^i | z_{1:t}, d_{1:t} \right),$$

(21)

where N_j is the number of particles for the relative pose. Here, before adding the relative pose to the particles of the camera path, we prune some hypotheses on the camera path on the basis of their weights. In our implementation, only the 10 best particles remain.

4.7 Landmark Estimation

For the efficient estimation of 3D landmarks, we employ a FastSLAM-type approach in which each 3D landmark is individually estimated with the given camera path.

We model the posterior distribution of each landmark $p(m_l | n_{1:k}, z_{1:k}^l, d_{1:k})$ defined in Eq. (12) using an optimized 3D landmark location, $\hat{m}_l$, and its uncertainty in the image space, e_l; we re-triangulate all observations including first two stereo views corresponding to each landmark for each particle by using SVD [18] to compute $\hat{m}_l$, and e_l is determined by the projection error of $\hat{m}_l$ for the last pose, n_N, as

$$e_l = \left\| z_N^l - z\left(\hat{m}_l, n_N \right) \right\|,$$

(22)

where N is the number of camera poses that observe the landmark.

5 Experimental Results

For experiments, we used a stereo camera with a 12 cm baseline and a 6 mm lens, which provide a narrow field of view.

5.1 Outdoor Experiments

Figure 7b and c shows the global maps and the camera paths computed by visual odometry and by using the proposed method with the corresponding Google map, respectively. For this experiment, we captured 10,667 images while walking more than 400 m in the outdoor environment with the stereo camera in hand. During this experiment, 325 key-frame images were designated from a sequence of images, which means that our key-frame SLAM only estimates the camera path that is

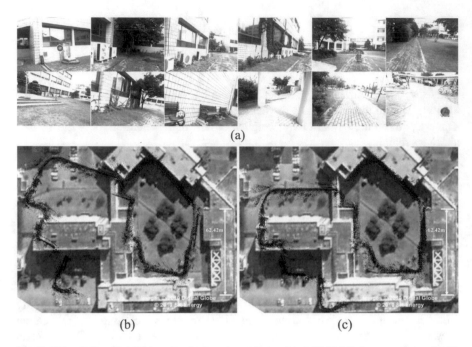

Fig. 7 The results obtained by visual odometry and our visual SLAM that are overlapped with the corresponding Google map (red dots: the key-frame locations of the camera, black points: the locations of the landmarks). (**a**) Some key-frame images designated during SLAM. (**b**) The top-down view of the 3D map and the key-frame locations estimated by visual odometry. (**c**) The top-down view of the 3D map and the key-frame locations estimated by using the proposed method

composed of 325 camera poses, and 110,076 landmarks were estimated. Figure 7a shows some key-frame images. We can see that the results obtained with the proposed visual SLAM approach, in which the visual odometry and stochastic filtering are combined, are much better than those obtained by only visual odometry.

Unfortunately, comparison with GPS data is indeed impossible due to unreliable outputs of the GPS, as shown in Fig. 2. Instead, we compared the proposed SLAM results with those of local bundle adjustment, because standard global bundle adjustment using all landmarks and the entire camera path cannot be feasibly performed for a long image sequence. A local bundle adjustment method ensures the good accuracy and consistency of the estimated camera poses as introduced in [24]. Recently, in [32], the authors analyzed the relative advantages of filtering and sparse optimization for sequential monocular SLAM.

Since we use a stereo camera, we include the feature coordinates in both left and right images in performing bundle adjustment to eliminate the depth ambiguity. In our dataset, we carried out local bundle adjustment when a new key frame is selected (the number of matched points with the last key frame is below 80% of the

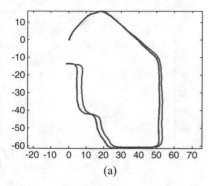
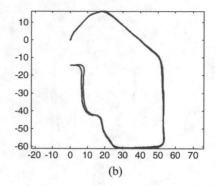

(a)   (b)

Fig. 8 Camera trajectories estimated by the proposed method (blue) and local bundle adjustment (red) along with different numbers of iterations (m). (**a**) Comparison with local bundle adjustment (15 LM iterations). (**b**) Comparison with local bundle adjustment (40 LM iterations)

observed features in the last key frame). For one local bundle adjustment, we used 10 last key frames and it is reported that local bundle adjustment can be performed within 73 ms when we set the number of Levenberg–Marquardt (LM) iterations to 15. Figure 8 shows the estimated paths obtained from visual SLAM and local bundle adjustment along with different numbers of LM iterations. As the number of iterations increases, the estimated path from local bundle adjustment is more similar with the path estimated from our visual SLAM.

In contrast to bundle adjustment, our visual SLAM recursively estimates the distribution of the state by using only incoming observations without previous observations and achieves the accuracy for the camera poses and landmarks as well as the computational efficiency. In addition, the part of stochastic filtering can be faster if it is implemented using multiprocessing programming such as Open Multi-Processing (OpenMP).

To show the validity of the proposed method that ensures the independence between process and measurement noise, we compute the camera path when we do not divide the observations. This means that all the observations are simultaneously used for both the process model and the measurement model. Figure 9a shows the results for this case. We observe that the results in Fig. 9a are less consistent than those of the proposed method that divides the observations according to each purpose, as shown in Fig. 8.

Our SLAM approach divides features into two subsets and uses the decreasing number of features for each model. To show the importance of achieving independence between two models, we intentionally removed some features for proposed SLAM and show the experimental results along with different numbers of features as shown in Fig. 9b–d that are more consistent than the result of Fig. 9a. These results demonstrate that achieving independence between process and measurement models is important even if it reduces the number of observations for each model.

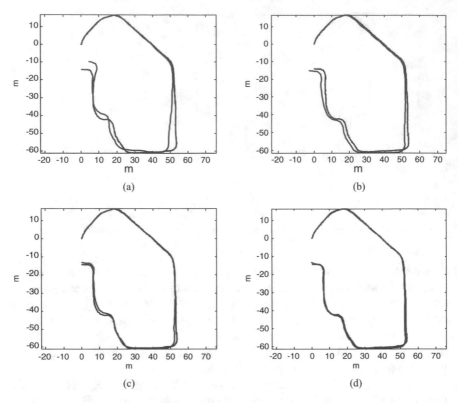

Fig. 9 The paths of visual SLAM along with different numbers of features (red: the camera path obtained from local bundle adjustment using 40 LM iterations, blue: the camera path of visual SLAM). (**a**) The camera path of visual SLAM when all features are used but those are not divided. (**b**) The camera path of visual SLAM when 70% of features are used and those are divided. (**c**) The camera path of visual SLAM when 80% of features are used and those are divided. (**d**) The camera path of visual SLAM when 90% of features are used and those are divided

Figure 10 shows the SLAM results for another environment after we captured an image sequence from a hand-held stereo camera in an outdoor environment while walking approximately 415 m and forming a loop. Finally, 390 key-frame poses and 95,125 landmarks were estimated. Figure 10 shows the estimated camera paths computed by visual odometry and by our visual SLAM.

To quantitatively measure the accuracy of our SLAM, we calculated the final pose errors and Table 1 lists the final pose errors for translation and rotation. For this purpose, we applied a corner matching method [33] to the third key-frame image and the last image which are influenced by view and illumination changes as shown in Fig. 11 because the KLT tracker cannot provide reliable correspondences between them.

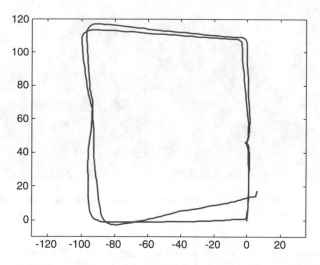

Fig. 10 The top-down view of the camera path (m) (blue: visual odometry, red: our visual SLAM)

Table 1 Final pose errors of three different methods for the outdoor experiment (approximately 415 m navigation) shown in Fig. 10

	Translation error			Rotation error		
	x-Axis	y-Axis	z-Axis	Roll	Pitch	Yaw
Visual odometry	4.049 m	8.381 m	14.469 m	2.305°	11.499°	1.193°
SLAM using only Key-frame measurements	0.091 m	6.790 m	2.014 m	3.497°	7.623°	1.134°
Proposed SLAM	1.199 m	0.122 m	0.935 m	0.784°	1.764°	1.117°

5.2 Evaluation on Various Conditions

The performance of our SLAM can be affected by various factors such as the number of particles, the criterion for key- frame selection, etc. In this section, we evaluate our SLAM performance on various conditions.

5.2.1 The Number of Particles

We run our SLAM algorithm while varying the number of particles for the path and the relative motion, and estimated paths are shown in Fig. 12. Table 2 shows the final pose errors with different numbers of particles. As we increase the number of particles for the relative motion, we can significantly improve SLAM performance compared to varying the number of particles for the path in that many particles are generated from our process model and our SLAM can accurately evaluate the probabilities of particles from our measurement model.

Fig. 11 Corner matching results between the third key-frame image and the last image for calculating final pose errors

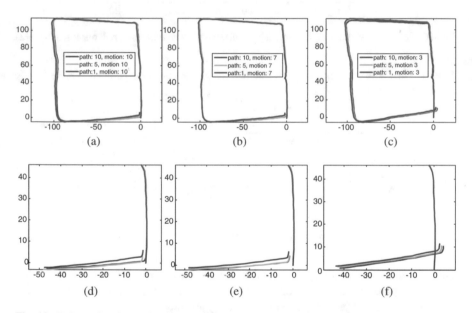

Fig. 12 Estimated paths obtained by the different number of particles from the outdoor experiment (approximately 415 m navigation) shown in Fig. 10. (**a**) 10 motion particles. (**b**) 7 motion particles. (**c**) 3 motion particles. (**d**) Enlarged results for **a**. (**e**) Enlarged results for **b**. (**f**) Enlarged results for **c**

Table 2 Final pose errors on different numbers of particles for the outdoor experiment (approximately 415 m navigation) shown in Fig. 10

	Translation error			Rotation error		
	1 particle	5 particles	10 particles	1 particle	5 particles	10 particles
3 motion particles	12.32 m	12.13 m	11.92 m	10.72°	9.01°	9.45°
7 motion particles	9.52 m	4.85 m	5.12 m	7.83°	3.92°	3.99°
10 motion particles	9.12 m	3.72 m	1.52 m	6.34°	2.41°	3.66°

5.2.2 The Criterion for Key-Frame Selection

We run our SLAM algorithm while changing the criterion for key- frame selection that the number of tracked features is below a certain percentage of the number of features in the last key frame. Thus we change the value of a percentage threshold between 30% and 70% to designate different key frames. Figure 13 shows the estimated paths from different percentage thresholds and Fig. 14 shows the final pose errors. As we increase the value of the percentage threshold, more key- frames are designated, consequently, more prediction steps are performed, which yields significantly accumulated error. On the contrary to this, we decrease the value of the percentage threshold, then our SLAM uses the decreasing number of measurements to update the camera path and the map, which also yields accumulated error. For this reason, 50% for the threshold value shows the best performance in this evaluation.

5.2.3 The Level of Independence

One of the contributions for our SLAM is to ensure the independence between the process noise and the measurement noise when using a single visual sensor. In order to demonstrate the importance of independency between the process and measurement noise, we first intentionally add some of new features, z_d, to a set of common features, z_c and calculate the final pose errors along with the different

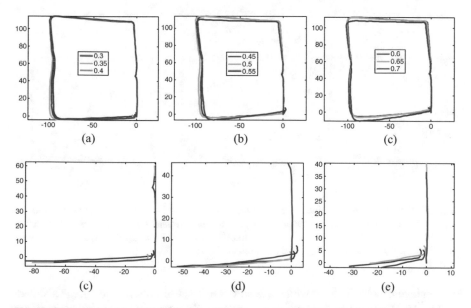

Fig. 13 Estimated paths obtained by the different criterion for key- frame selection from the outdoor experiment (approximately 415 m navigation) shown in Fig. 10. (**a**) From 0.3 to 0.4. (**b**) From 0.45 to 0.55. (**c**) From 0.6 to 0.7. (**d**) Enlarged results for **a**. (**e**) Enlarged results for **b**. (**f**) Enlarged results for **c**

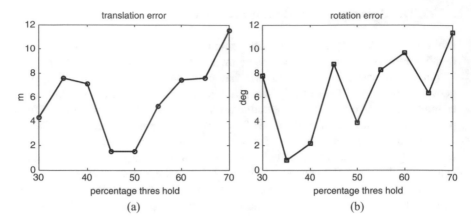

Fig. 14 Final pose errors for the different criterion for key- frame selection from the outdoor experiment (approximately 415 m navigation) shown in Fig. 10. (**a**) Translation error. (**b**) Rotation error

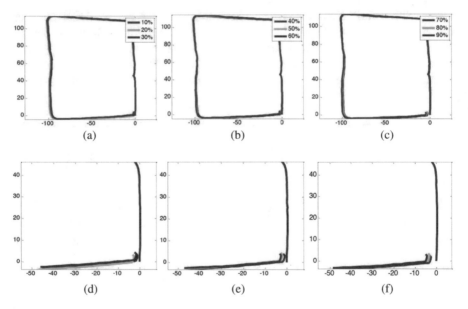

Fig. 15 Estimated paths obtained by adding the different percentange of new features, z_d to the common feature set z_c from the outdoor experiment (approximately 415 m navigation) shown in Fig. 10. (**a**) From 10% to 30%. (**b**) From 40% to 60%. (**c**) From 70% to 90%. (**d**) Enlarged results for **a**. (**e**) Enlarged results for **b**. (**f**) Enlarged results for **c**

percentage of new features which are added to a set of common features as shown in Fig. 15. We show the final pose errors with different percentage of added features in Fig. 16. When we perfectly decouple visual features into two sets, the proposed SLAM algorithm shows the best performance for the outdoor dataset.

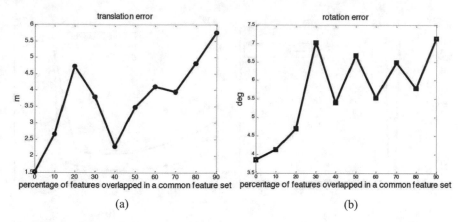

Fig. 16 Final pose errors for the different percentage of new features z_d in the common feature set z_c from the outdoor experiment (approximately 415 m navigation) shown in Fig. 10. (**a**) Translation error. (**b**) Rotation error

5.3 Evaluation on Karlsruhe Datasets

To evaluate the proposed SLAM performance, we also used Karlsruhe datasets[2] which contain stereo sequences recorded from a moving vehicle in Karlsruhe and corresponding ground truth coordinates from a GPS sensor. These sequences were used in [34] as well.

For evaluation, we tested our SLAM approach by using the first four stereo sequences in Karlsruhe datasets; *2009-09-08-drive-10*, *2009-09-08-drive-12*, *2009-09-08-drive-15*, and *2009-09-08-drive-16* sequences. We evaluated the accuracy of the estimated paths using the GPS coordinates as shown in Fig. 17. To clearly show the error accumulation problem when sequentially performing visual SLAM using all the frames,[3] we additionally show the camera paths computed from sequential SLAM in Fig. 17. According to our results shown in Fig. 17, our key-frame SLAM approach reduces the number of predictions, which increase the uncertainties, by reducing the number of camera poses to be estimated in the path, and updates the distribution of the camera path by using many measurements obtained at the non-key-frame locations to further decrease the uncertainties.

Table 3 lists the numbers of key frames and landmarks estimated by our visual SLAM for each stereo sequence and shows the location errors computed by GPS coordinates.

Figure 18 shows camera paths and maps from four stereo sequences estimated by using the proposed SLAM approach.

---

[2]Datasets are available at http://www.cvlibs.net/datasets.html.

[3]For this purpose, we designate all the frames as key frames, and the camera path consists of the camera poses for all the frames.

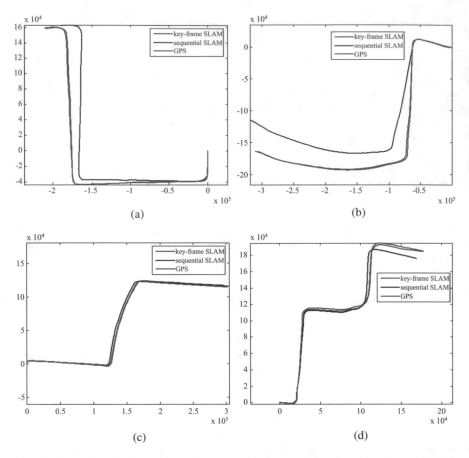

Fig. 17 Comparison between estimated camera paths of proposed key-frame-based visual SLAM, sequential visual SLAM, and GPS coordinates for Karlsruhe datasets (mm). (**a**) The camera paths with GPS data for the 2009-09-08-drive-0010 stereo sequence (432 m). (**b**) The camera paths with GPS data for the 2009-09-08-drive-0012 stereo sequence (488 m). (**c**) The camera paths with GPS data for the 2009-09-08-drive-0015 stereo sequence (395 m). (**d**) The camera paths with GPS data for the 2009-09-08-drive-0016 stereo sequence (344 m)

Table 3 Evaluation of our SLAM results on Karlsruhe datasets

Sequence	Number of frames	Number of key frames	Number of landmarks	Total distance (m)	Final location error (m)	Maximum error (m)
drive-10	1423	114	20,197	432	1.2943	1.7723
drive-12	910	141	30,806	488	4.4719	4.8532
drive-15	1021	127	21,003	395	3.3865	3.3865
drive-16	1206	162	48,065	344	1.5475	2.5006

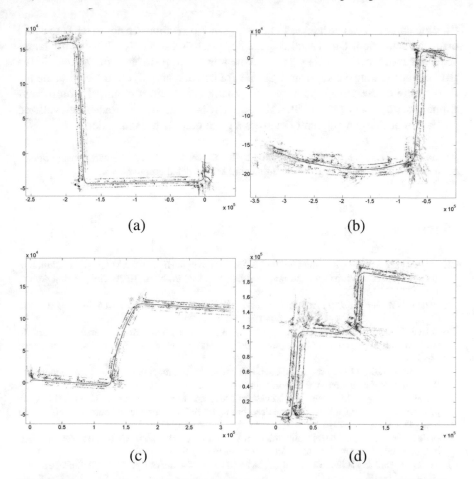

Fig. 18 The results of our visual SLAM for Karlsruhe stereo sequences (mm). (**a**) The camera path and the map using the 2009-09-08-drive-0010 stereo sequence. (**b**) The camera path and the map using the 2009-09-08-drive-0012 stereo sequence. (**c**) The camera path and the map using the 2009-09-08-drive-0015 stereo sequence. (**d**) The camera path and the map using the 2009-09-08-drive-0016 stereo sequence

6 Conclusion

We presented a novel visual SLAM method in which visual odometry and key-frame-based Bayesian filtering are combined to cope with sudden camera motion and to obtain consistent maps. Visual odometry and Bayesian filtering compensate for each other's drawbacks—our visual SLAM approach can make robust predictions for the camera pose owing to visual odometry, while proposed key-frame-based Bayesian filtering alleviates error accumulation involved in visual odometry. In order to ensure the independence between the process and measurement noise, we divide observations into two categories. The proposed Bayesian filtering approach can be adopted in existing motion estimation approaches to avoid error accumulation. Our visual SLAM approach is especially efficient for large-scale environments since (1) we reduce the number of possible camera poses in the path by formulating the key-frame SLAM framework, and (2) effectively update the distribution of the camera path by using many observations obtained at non- key-frame locations, and (3) our SLAM approach is based on the Rao-Blackwellized particle filter that can efficiently manage a great number of landmarks.

Acknowledgements This work is supported by the Center for Integrated Smart Sensors funded by the Ministry of Science, ICT & Future Planning as the Global Frontier Project.

References

1. Dissanayake MWMG, Newman P, Clark S, Durrant-Whyte HF, Csorba M (2001) A solution to the simultaneous localization and map building (SLAM) problem. IEEE Trans Robot Autom 17(3):229–241
2. Welch G, Bishop G (2001) An introduction to the Kalman filter. In: SIGGRAPH 2001 (Course 8)
3. Torma P, György A, Szepesvári C (2010) A Markov-Chain Monte Carlo approach to simultaneous localization and mapping. In: International conference on artificial intelligence and statistics
4. Davison AJ (2003) Real-time simultaneous localisation and mapping with a single camera. In: IEEE international conference on computer vision, vol 2, pp 1403–1410
5. Chekhlov D, Pupilli M, Mayol W, Calway A (2007) Robust real-time visual SLAM using scale prediction and exemplar based feature description. In: IEEE conference on computer vision and pattern recognition, pp 1–7
6. Eade E, Drummond T (2007) Monocular SLAM as a graph of coalesced observations. In: IEEE international conference on computer vision, pp 1–8
7. Kaess M, Dellaert F (2006) Visual SLAM with a multi-camera rig. GVU Technical Report, GIT-GVU-06-06
8. Marks TK, Howard A, Bajracharya M, Cottrell GW, Matthies L (2008) Gamma-SLAM: using stereo vision and variance grid maps for SLAM in unstructured environments. In: IEEE international conference on robotics and automation, pp 3717–3724
9. Montemerlo M, Whittaker W, Thrun S (2002) FastSLAM: a factored solution to the simultaneous localization and mapping problem. In: Proceedings of the AAAI national conference on artificial intelligence, pp 593–598
10. Davison AJ, Reid ID, Molton N, Stasse O (2007) MonoSLAM: real-time single camera SLAM. IEEE Trans Pattern Anal Mach Intell 29(6):1052–1067
11. Doucet A, Freitas N, Murphy K, Russell S (2000) Rao-bBlackwellized particle filtering for dynamic Bayesian networks. In: Proceedings of the 16th conference on uncertainty in artificial intelligence
12. Doucet A, Freitas N, Gordon N (2001) Sequential Monte Carlo methods in practice. Springer, New York
13. Kim C, Sakthivel R, Chung WK (2008) Unscented FastSLAM: a robust and efficient solution to the SLAM problem. IEEE Trans Robot 24(4):808–820
14. Eade E, Drummond T (2006) Scalable monocular SLAM. In: IEEE conference on computer vision and pattern recognition, vol 1, pp 469–476
15. Julier S, Uhlmann J, Durrant-Whyte HF (2000) A new method for the nonlinear transformation of means and covariances in filters and estimators. IEEE Trans Autom Control 45(3):477–482

16. Paz LM, Pinié P, Tardós JD, Neira J (2008) Large-scale 6-DOF SLAM with stereo-in-hand. IEEE Trans Robot 24(5):946–957
17. Pupilli M, Calway A (2006) Real-time visual SLAM with resilience to erratic motion. In: IEEE conference on computer vision and pattern recognition, vol 1, pp 1244–1249
18. Hartley R, Zisserman A (2000) Multiple view geometry in computer vision. Cambridge University Press, Cambridge
19. Nister D, Naroditsky O, Bergen J (2004) Visual odometry. In: IEEE conference on computer vision and pattern recognition, vol 1, pp 652–659
20. Zhu Z, Oskiper T, Samarasekera S, Kumar R, Sawhney HS (2008) Real-time global localization with a pre-built visual landmarks database. In: IEEE conference on computer vision and pattern recognition, pp 1–8
21. Triggs B, McLauchlan P, Hartley R, Fitzgibbon A (2000) Bundle adjustment - a modern synthesis. LNCS (vision algorithms: theory and practice). Springer, Heidelberg
22. Engels C, Stewénius H, Nistér D (2006) Bundle adjustment rules. In: Photogrammetric computer vision
23. Jeong Y, Nister D, Steedly D, Szeliski R, Kweon IS (2010) Pushing the envelope of modern methods for bundle adjustment. In: IEEE conference on computer vision and pattern recognition
24. Mouragnon E, Lhuillier M, Dhome M, Dekeyser F, Sayd P (2006) Real time localization and 3D reconstruction. In: IEEE conference on computer vision and pattern recognition, pp 363–370
25. Eudes A, Lhuillier M (2009) Error propagations for local bundle adjustment. In: IEEE conference on computer vision and pattern recognition, pp 2411–2418
26. Royer E, Lhuillier M, Dhome M, Lavest J-M (2007) Monocular vision for mobile robot localization and autonomous navigation. Int J Comput Vis 74(3):237–260
27. Konolige K, Agrawal M (2008) FrameSLAM: from bundle adjustment to real-time visual mappping. IEEE Trans Robot 24(5):1066–1077
28. Shi J, Tomasi C (1994) Good features to track. In: IEEE conference on computer vision and pattern recognition
29. Haralick R, Lee C, Ottenberg K, Nölle M (1994) review and analysis of solutions of the three point perspective pose estimation problem. Int J Comput Vis 13(3):331–356
30. Fischler M, Bolles R (1981) Random sample consensus: a paradigm for model fitting with application to image analysis and automated cartography. Commun ACM 24(6):381–395
31. Pollefeys M, Gool LV, Vergauwen M, Cornelis K, Verbiest F, Tops J (2002) Video-to-3D. In: Proceedings of photogrammetric computer vision 2002 (ISPRS commission III symposium), international archive of photogrammetry and remote sensing
32. Strasdat H, Montiel JMM, Davison AJ (2010) Real-time monocular SLAM: Why filter?. In: IEEE international conference on robotics and automation, pp 2657–2664
33. Kim J, Kweon IS (2010) Vision-based navigation with pose recovery under visual occlusion and kidnapping. In: IEEE international conference on robotics and automation, pp 1921–1927
34. Kitt B, Geiger A, Lategahn H (2010) Visual odometry based on stereo image sequences with RANSAC-based outlier rejection scheme. In: IEEE intelligent vehicles symposium

LED-Based Optical Neural Implants

Sunghyun Yoo, Sang Beom Jun, and Chang-Hyeon Ji

1 Introduction

For several decades, there have been a number of attempts to monitor and control the signaling in the nerve systems for the purposes of investigation of brain functions as well as treatment of neurological diseases. Traditionally, electrical methods have been employed for both the detection of neural activity and the stimulation of the nerves. For example, a single neuron can be stimulated, and the action potentials from the neuron can be recorded via patch-clamp methods for research purposes [1]. On the other hand, a population of neurons in the nerve system can also be stimulated or recorded noninvasively via electroencephalogram and transcranial direct current stimulation [2–5]. Due to the remarkable advances in electronic systems, the electrical communication with nerve systems has become one of the most conventional methods. The microfabrication technology also enabled the development of microelectrode arrays which can provide high-density electrode-neuron interfaces for in vivo and in vitro applications [6, 7]. Nowadays, the strategy of electrical stimulation on nerve system is not only applied for research but also successfully applied to treat neurological disorders as well as to restore the disabled sensory functions such as vision and hearing [8–10]. Deep brain stimulation system has been successfully used for more than a decade to treat a variety of neurological diseases such as Parkinson's disease, essential tremor, and dystonia via delivering electrical voltage pulses to the target brain area through implanted electrode [11,

S. Yoo
Department of Electrical and Computer Engineering, Seoul National University, Seoul, Republic of Korea

S. B. Jun · C.-H. Ji (✉)
Department of Electronic and Electrical Engineering, Ewha Womans University, Seoul, Republic of Korea
e-mail: cji@ewha.ac.kr

© Springer Nature Switzerland AG 2020
Y. Liu et al. (eds.), *Smart Sensors and Systems*,
https://doi.org/10.1007/978-3-030-42234-9_3

12]. Auditory prosthesis system had been developed more than 30 years ago and has enabled the restoration of hearing ability in the people with profound deafness [13, 14]. Recently, based on the similar design principle, visual prosthetic system is also developed and clinically applied to people with blindness [15, 16].

Despite the successful outcome of electrical approaches to connect to nerve systems in both researches and clinical applications, there are still several drawbacks. For example, the electrical signals recorded from individual neurons or a group of neurons include all the signals from every electrical sources neighboring the electrode without any information of neuronal cell types. However, as the extensive knowledge in neuroscience has accumulated, it is becoming more important to identify the signaling from specific types of neurons because different cell types form different synapses via a specific neurotransmitter–receptor pair. Similarly, electrical stimulation also has a limitation of nonspecific targeting of neuronal cells because ionic current flow diffuses away only depending on the conductive path. Therefore, several researchers have raised the need for different approaches to detect and modulate the activity of specific neuronal cell types.

In order to overcome the drawback of electrical neural interfaces, various methods have been proposed for neural interfaces including optical, mechanical, thermal, chemical, and magnetic neural stimulation and/or detection. It is because there have been several new findings regarding the mechanisms for activation or blocking of neural signaling. A number of different alterations of surrounding conditions such as optical irradiation, mechanical vibration, temperature change, and chemicals binding to receptors are shown to be effective in modulating the activity of neurons. Focused magnetic field is also employed to induce the activation of neurons via electromagnetic coupling in the electrolytes and now being used for clinical treatment. Among those various techniques for neural modulation, during the last decade, optogenetics has become one of the most powerful research tools to selectively control or modulate the activity of specific types of neurons, which has been impossible with conventional electrical methods. As one of optical neural interface methods, optogenetics has been enabled by the use of exogenous optical indicators or actuator mostly based on genetically expressed fluorescence proteins.

For several decades, the use of fluorescence protein was utilized mainly for the purpose of fluorescent microscopy. Afterward, due to the development of genetic techniques such as gene modification through viral vector injection or cell transfections, the fluorescence optics has become a unique tool to identify specific subcellular structures. In addition, combining fluorescence proteins and genetics finally enabled not only the morphological imaging but also the novel powerful tools for neural interface. The fundamental mechanism is to optically control or detect the activity of genetically modified neurons in the nerve system simply by exposure to a light of specific wavelength, which is called optogenetics. More specifically, optogenetics is understood as a technique to control the light-sensitive ion channels expressed on genetically selected nervous tissues. In addition to the modulation of channel currents, the activity-dependent fluorescence markers can give us a new way to optically detect the activity of genetically selected neurons. For instance, the development of genetically encoded Ca^{2+} activity-dependent fluorescence has provided a new paradigm of neuroscience research, which enabled the simultaneous monitoring of hundreds of neurons through fluorescence microscopy [17–19].

Even though the optogenetic techniques have become an important tool in neuroscience field, there are still several hurdles to be overcome to be applied to animal studies or clinical applications. Compared to traditional electrical neural interfaces, the optical equipment is difficult to be miniaturized due to high power consumption, bulky light sources, coupling between light source and waveguides, and so on. Therefore, during the last decade, there have been several attempts to develop a miniaturized, implantable, power-efficient, multifunctional optical systems for the neural interfaces and mainly for optogenetic applications. For example, head-mountable and miniaturized optical instruments were developed so that neuronal activities in the brain can be measured while the experimental animal can freely move around and perform specific behavior tasks [20–27].

This chapter describes the current state-of-the-art technologies for implantable optical neural interfaces based on the optogenetic technology mostly for neuromodulation. The implantable system comprises of several components and tools, which include implantable waveguides, light sources, control electronics, power supply, tools for implantations and combination with electrical interfaces, and so on. LED-based optical interface systems are mostly described in this chapter because the use of laser and conventional optics is inappropriate for a miniaturized and wireless implantable systems for animal or clinical applications. Two different types of LED-based systems are compared. First approach is to use an LED for direct light illumination of the target nerve cells, and the other is to use an LED coupled with waveguides. Several approaches with the waveguides are also compared in terms of efficiency, fabrication complexity, multifunctionality, and biosafety. In addition to current relevant technologies, existing challenges are discussed to achieve the ideal implantable optical system for practical purposes.

2 Optical Neural Interfaces

Researches on optogenetics up to the present can be categorized into two main branches: the development of opsin genes including transgenic technology [28] and the development of the optical neural interfaces [29]. Progresses in either branch to improve the opsins or the control interfaces are expected to extend the scope and application of experiments which will significantly contribute to the broadening of human knowledge of the neural system as well as to the treatment of intractable neurological diseases.

Implantable optical neural interface for modulating nerve functions can be defined as a miniaturized device that delivers light with specific wavelengths to the targeted area of the neural system. During the past decade, there has been a significant progress in the development of these types of devices, which include the integration of multiple sensing electrodes, microfluidic channels for drug injection, and temperature monitoring sensors. In general, implantable optical neural interface can be grouped into two major types depending on the type of light sources used: laser-based [30–41] and LED-based systems [42–47].

At the early stage of the development, the optical fibers coupled with external laser sources were the most common tools for light delivery to targeted neural cells. A bare optical fiber was directly utilized as an optical waveguide for direct light delivery to a living rodent's target region in the brain for the first time in an experiment conducted by Dr. Deisseroth and his group at Stanford University [30, 31]. Figure 1 illustrates the typical optical fiber-based light delivery probe and operation principle. This approach has been one of the most widely used methods for the past decade since the researchers without in-depth knowledge of optics can easily construct an experimental setup by combining commercially available optical components.

Among various light sources, laser has many advantages over other candidates in terms of high output light intensity, low light beam divergence, and coherence with narrow bandwidth. Direct illumination of neural cells via optical fiber fully utilizes the advantages of fiber-optic and laser-based systems while maintaining a low light loss performance. This approach also benefits from the availability of various mature peripheral equipment technologies used in the field of fiber optics. Moreover, compared to other types of approaches, direct illumination using a bare optical fiber allows for more scalability in probe length, which is critical for the stimulation of deep brain regions. It has also been shown that the laser-based optical fiber systems can be integrated with metal electrodes for neural signal recording [32, 33], polymer-based waveguides [34, 35], and glass optrode array [36]. Examples of optical neural interfaces utilizing laser light sources are summarized in Fig. 2.

Despite the numerous advantages mentioned above, conventional fiber-optic probe combined with a laser source has fundamental limitations. It is very challenging to realize an untethered and miniaturized system with wireless control capability using this configuration, as a complicated and bulky structure is inevitable for the

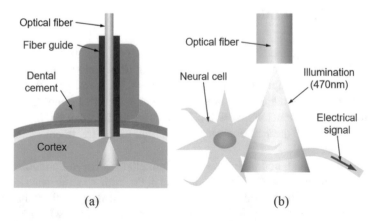

(a)   (b)

Fig. 1 Optical fiber as a light delivery probe: (**a**) typical implanted optical fiber for optogenetics and (**b**) illustration of optical neural cell stimulation and electrical signal generation [31]

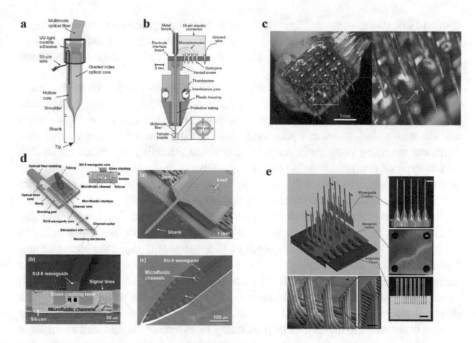

Fig. 2 Various optical neural interfaces utilizing laser light sources [37–41]: (**a**) dual-core optical fiber system for both optical stimulation and electrical recording, (**b**) multimode optical fiber assembled with four tetrode bundles for recording, (**c**) optrode array with a multimode optical fiber adapted from a Utah electrode array, (**d**) in-plane neural probe with embedded waveguide and microfluidic channels, (**e**) waveguide array with waveguide combs assembled on a base plate holder (Reproduced from Ref. [48] with permission from The Royal Society of Chemistry)

coupling between the laser light source and optical fiber. The need for an untethered system arises from the growing demand for neuroscience experiments using freely behaving animals. Moreover, the challenge becomes more severe when multiple optical fibers or an array of fibers are required for increased exposure area or integration with additional electrodes is required for reading out electrical signal. Limited scalability of the fiber array and low spatial resolution of the fibers are added issues that need to be addressed in these types of fiber–optic probe system, considering that the fibers need to be glued and fixed to the bulky guide structure.

Considering the recent demand for the miniaturization of optical neural interfaces, utilization of LED as a light source, instead of the laser counterpart, can be an option. Since the size of an LED can be less than a millimeter on one side, ranging from tens to hundreds of micrometers, various integration approaches can be utilized with the aid of microfabrication technology, which resulted in a successful demonstration of light delivery probes integrated with micro LEDs (μLEDs) [42, 49]. Availability of commercial LED products with wide variety of specifications in terms of size, wavelength, output power, and configuration is another advantage [50–52]. LEDs typically have small size with high spatial resolution and low power consumption, which are essential properties for implementing a wirelessly controlled implantable optical neural interface system. In addition, stable illumination and fast switching speed also make LEDs suitable for the optical neural interfaces.

In contrast to the advantages mentioned above, LED-based probes also have drawbacks. Due to the incoherence and Lambertian light emission, LED-based probes inherently suffer from low light coupling efficiency between the light source and the waveguide, which can potentially lead to light power handling issue. Also, concerns are being posed, which are related to the reliability and biocompatibility of this type of configuration due to the high amount of localized current and heat generated during operation. Still, direct cell illumination using μLEDs gives more efficient light emission than illumination through waveguides.

Recently, significant research has been conducted on both the laser-based and LED-based optical neural interfaces to achieve efficient light delivery together with integration of other functionalities, such as bidirectionality (simultaneous stimulation and response signal sensing), multiwavelength illumination, drug delivery, and various sensing capabilities. Both methods have respective advantages and disadvantages, as well as relevant requirements, depending on the specific applications where the developed system will be used. Although many studies are still under way to overcome the limitations of previously reported approaches and to improve the performance of individual systems and provide unprecedented capabilities, LED-based probes can potentially be advantageous for the implementation of an untethered system and are being utilized more in researches targeting the realization of implantable optical neural interfaces.

3   LED-Based Light Probe

LED-based light delivery probes can be categorized based on the types of LEDs utilized, which are commercially available LED chips and microfabricated μLEDs (Fig. 3). For the commercially available LED chips, a number of manufacturers such as Cree, Osram, Kingbright, and SunLED offer packaged LED chips that are only a few millimeters in size. Some of these chips are available in multicolor, which can be utilized in the development of the multiwavelength optical neural interfaces. The luminescence characteristics and the power efficiency of the commercial LED chips have been improving steadily. Due to the significant growth of the μLED-based display industry, μLEDs using III–V compound semiconductor are rapidly becoming much smaller in size than the conventional LEDs [53], and the optical and the electrical performances are also being improved dramatically. These advancements in LED technology are expected to have a considerable impact on the development of LED-based neural probes.

The LED-based neural probes for implantable systems have been proposed and developed through a wide variety of approaches, each of which has contributed to the advances in the field of optogenetics with their unique and excellent

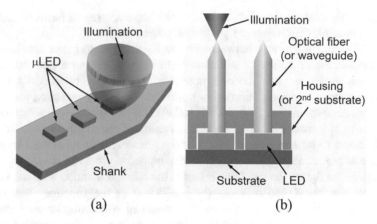

Fig. 3 Typical approaches with LED-based light probes: (**a**) commercial LED-based probe and (**b**) μLED-based probe

performances. This subchapter focuses on the light delivery for neuromodulation, bidirectionality, reliability, biocompatibility, and multifunctionality of LED-based light probes.

3.1 Light Delivery Using LED for Neurostimulation

Efficient delivery of sufficient light power is one of the most important and fundamental requirements for optical neural interfaces for optogenetics. Effective light delivery becomes more crucial when the probes are designed for a wirelessly controlled implantable system with limited electrical energy, which is becoming more common in these days. Although wireless supply of electrical power can be considered, many of the wireless neural interfaces under development are based on integrated energy storage components. For effective delivery of light to the nerve system, light intensity, luminous area, and coupling efficiency between the optical components in the probe should be taken into account.

The light intensity required for the activation of light-sensitive opsins is well known to be above $1-5$ mW/mm^2 for in vitro stimulation with blue light [54, 55] and over 7 mW/mm^2 for in vivo inhibition when measured with red light [56]. However, simply estimated light intensity under conventional laboratory environment is often not enough in practice due to light absorption and scattering in tissues. In principle, the light intensity is attenuated exponentially as a function of penetration depth [30]. It has been reported that the irradiance decreases to below 10% of the value measured at the tissue surface within approximately 300 μm depth [30, 57]. Therefore, practical in vivo photostimulation of living animals requires a much higher grade of light power at the probe tip. Alternatively, an accurate insertion that puts the probe facet as close as possible to the targeted area in terms of insertion depth can minimize the required light output power.

Precise lateral alignment between the probe and the targeted area can also affect the success or failure of sufficient light delivery during a practical animal experiment. Figure 4 plots the maximum instantaneous light intensity used in photostimulations on living animals as a function of the illumination area of the probe from three laser-based [34, 35, 58] and six LED-based [42–47] neural probes. While the light intensity required for the activation of channelrhodopsin-2 (ChR2) is well known to be 1 mW/mm^2 [54, 55], the practical intensities used in animal experiments are required to be several orders higher, up to 1714 mW/mm^2, at the probe tip. In the figure, the solid line refers to the rational function with the average light output power (0.946 mW) as the coefficient of the function. Distribution of the data points in Fig. 4 indicates that practical photostimulation requires a certain amount of total light power. Furthermore, the optical probes with a smaller illumination area require a relatively higher light intensity in practice than the probes with larger illumination area, as the probes with small illumination area are more vulnerable to lateral misalignment. However, since there is a risk of photo damage of overexposed tissue, there also exists a limit for the increased light intensity. Therefore, according to the desired three-dimensional illuminated brain area, the intensity and the illumination property of light should be determined carefully.

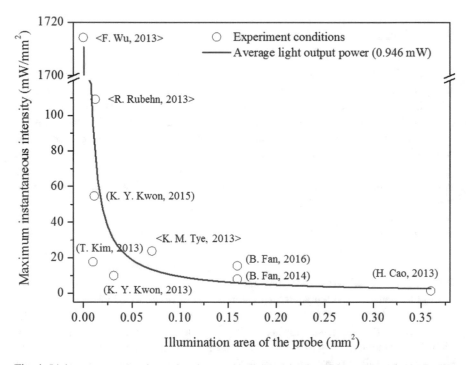

Fig. 4 Light output power of neural probes used in living animal experiment [34, 35, 42–47, 58]

Precise measurement of the light intensity in the three-dimensional space filled with material having optical properties close to tissue or accurate simulation of light distribution using ray tracing tool can be helpful.

Despite the practical requirement of total light power for photostimulations on living animals, current LED-based neural probes with integrated waveguides generally have low power coupling efficiency, which is the amount of light power delivered at the probe tip compared to the consumed electrical power. Figure 5 summarizes the power coupling efficiency of current LED-based neural probes [42–44, 46, 47, 59, 60]. It clearly verifies that the probes with waveguides have much lower average power efficiency of −27.6 dB when compared with the probes with direct LED exposure. In this sense, LED-based probes with waveguides require a power handling capability of 544 mW in order to achieve 0.946 mW output light power, which is relatively high for untethered implantable systems, especially for biomedical applications. On the other hand, LED-based probes without waveguides require only 61 mW for the same output light power. Due to the low light coupling efficiency between the LED and the waveguide, direct utilization of μLED is more competitive for an implantable system, in terms of power handling issue although

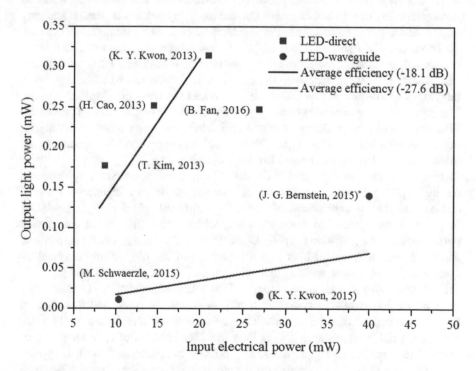

*: Values were scaled down to 1/50 (actual input and output power are 2 W and 7 mW, respectively.)

Fig. 5 Comparison of power efficiency for LED-based neural probes [42–44, 46, 47, 59, 60]

other issues such as packaging, heat dissipation, and biocompatibility arise when the μLED is to be implanted in the nerve tissue.

3.2   Reliability and Biocompatibility

Ensuring both reliability and biocompatibility is the most critical requisite of implantable neural interfaces for biomedical applications. Many researchers have provided well-organized validation of safety matters in their devices, as probes are generally inserted inside the living body with direct contact to neural cells. Direct insertion of a μLED into the brain still requires utmost care with regard to both localized heat dissipation and electrical leakage. Several researches have suggested integrating either a polymer waveguide or optical fibers as possible solutions to the potential hazards of the direct μLED insertion approach [46, 50, 59, 60]. Integration of the LED chip and optical fiber in the macroscale is a proven technology with relatively simple and straightforward process compared to the integration of μLED on a microfabricated probe tip. However, the use of optical fibers as waveguides involves a labor-intensive manual assembly process in general and may lead to difficulties in ensuring a uniform performance among manufactured devices. To overcome the issue with assembly process, which can be critical for mass production, a standardized fabrication approach or utilization of microfabricated components is required. However, the use of optical fibers can be advantageous in fabricating a two-dimensional (2D) array of probes to increase the illumination area. The polymer waveguides are generally fabricated using microfabrication processes. Therefore, it can be precisely designed and fabricated to meet the requirements of the user and has excellent expandability such as integration with electrodes for readout and a microfluidic channel for drug injection.

In terms of the materials used in the device, polymers are not yet fully validated for its long-term biocompatibility, which can potentially be a significant issue in chronic and clinical applications. Moreover, a significant light delivery degradation in polymer waveguide has been reported, which could also affect the device performance in the long term [45, 46]. Optical fiber has relatively small degradation, and the probes with optical fiber as a waveguide [50, 59, 60] are implemented by integrating a segmented optical fiber with an LED chip. Although not yet fully validated, existing studies have elaborated on the feasibility of a 2D fiber array for wirelessly controllable system [59, 60]. A wirelessly controlled illumination through a single ferruled fiber has also been demonstrated successfully [50]. Since the LED is significantly larger than the fiber in terms of light receiving or transmitting area in these approaches, a monolithic integration approach is applied for the fibers and the light source to maximize the light coupling efficiency between the LED and the fiber by arranging them in direct contact [61] and to free them from the alignment error during the assembly.

3.3 Bidirectionality

In general, the experimental procedure in optogenetics consists of optical control and probing. A light delivery probe is inserted into the body, and light is flashed on the neural cells, which draws the desired responses from only the targeted neurons through light-sensitive ion channels or pumps. The result of the photostimulation is analyzed by either observing the animal behavior or by measuring the electrical activity near the light-exposed neural cell.

In order to accurately examine the light-evoked neuronal activities, an electrical potential difference of several tens of microvolts or less around the cell should be measured. Therefore, in electrode design and arrangement for electrical recording, it is important to consider factors such as spatial proximity to the light illuminated region, high spatial resolution, simultaneous multiple spot recording, and impedance control of the electrode. As the electrical neurostimulation and recording have already been studied for decades ahead of optogenetics and photostimulation using the bare optical fiber has been demonstrated successfully, various approaches for the integration of the recording electrode with optical probe have been reported [62–64]. Examples include direct deposition of metal layer on optical fiber [62], integration of microfabricated silicon-based electrode array using UV curable epoxy [63], and flexibly arranging multiple optical fibers and electrodes using customized guide module, namely, the flexDrive [64].

Studies on the integration of such electrodes have also been actively conducted in LED-based optical light delivery devices [42, 43, 46, 47]. In these approaches, electrodes are formed with materials such as platinum, gold/chromium, and copper. Passivation layers are formed with biocompatible insulating materials for improved electrical properties and biosafety. Considering the ease of manufacturing process, mechanical robustness and flexibility, and heat dissipation, shank was fabricated with various materials such as polyethylene terephthalate (PET) [42], polyimide [43], polydimethylsiloxane (PDMS) [46], and polycrystalline diamond (PCD) [47].

3.4 Multifunctionality

Ever since the first validation on a living animal [30], optical neural probes have been improved in various aspects, which include the integration of functionalities such as bidirectionality, drug delivery capability, temperature sensing, and multi-wavelength illumination. Among these, multiwavelength illumination is essential for selective stimulation and inhibition using single device, considering that opsins such as channelrhodopsin (ChR) and halorhodopsin (HR) respond to lights of different wavelengths. Direct utilization of optical fiber and laser source allows for a relatively easy implementation of multiwavelength illumination. One way to realize the system is to simply use a 1:2 fiber coupler and two separate laser sources of different wavelengths. However, laser sources can be costly, especially for the yellow light.

Several hurdles must be overcome for an LED-based light delivery systems, in terms of probes both with and without the waveguides. In general, LED-based neural probes have issues such as low light coupling efficiency and low output light power. Therefore, implementing a multiwavelength illumination capability by integrating separate LEDs of different colors will make the situation even more complicated. This could potentially be the reason why there have been no reports, to the authors' knowledge, on multiwavelength LED-based light delivery probes with high perfection.

One approach to achieve a multiwavelength direct µLED illumination system is to place separate µLEDs on the substrate or the silicon-based tips. However, it is very difficult to achieve precise alignment and consistent illumination on the same target area in this type of arrangement. Moreover, extremely complicated and difficult fabrication procedure of installing multiple µLEDs is frequently mentioned as its drawbacks. Compared to direct illumination, probes with waveguides are even more vulnerable to light power handling issues. Conventional butting method is adopted in the probes utilizing optical fibers as waveguides to optimize the light coupling efficiency between the LED and the optical fiber. In this configuration, installing multiple LED sources in the system requires a novel approach to overcome the difficulties related with LED integration and alignment with the fiber.

In terms of the safety of animal during the experiment, heat generated by light illumination and probe circuitry is a critical issue. Temperature changes of approximately 2–3°C in neural system affect not only the outcome of the experiment but also the safety of the subject [65]. Therefore, many researchers are trying to maintain the temperature change of the distal end of the neural interface and its periphery less than 1°C during the experiment. In this context, it would be helpful to integrate a sensor that monitors the temperature change around the target tissues in real time during the experiment. Systems with temperature monitoring capability have been reported, but the performance is still limited due to difficulties with fabrication process and complexities of the fabricated system [42, 66].

4 Discussion and Conclusion

Table 1 shows the representative LED-based optical neural interfaces summarized in terms of advantages and current issues in the field of research. LED-based neural implants for optogenetics can be classified into two representative types: direct µLED-illuminated devices and probes with LED light sources and waveguides. Optical probes with waveguide structures employ either microfabricated polymer waveguides or commercially available optical fiber as waveguide.

While the practical instantaneous light power required for the optogenetic modulation is approximately 1 mW, current LED-based neural probes lack appropriate light power handling capability. In terms of sufficient light delivery, direct illumination using µLEDs can be preferable for systems targeting wirelessly controlled, implantable devices. On the contrary, probes with waveguides can be

Table 1 LED-based optical neural interfaces

	Typical devices and descriptions	Reference
Systems without waveguides	1×4 array of GaN μLED (6.45 μm thick, 50×50 μm^2), Pt electrode, Si photodiode, and temperature sensor stacked on an epoxy microneedle	[42]
Systems with waveguides	1×5 array of 40 μm diameter μLEDs fabricated from a commercial GaN 450 nm LED wafer with sapphire substrate	[67]
	Two 4×4 arrays of μLEDs integrated on a polyimide substrate bonded with array of SU-8 microwaveguides with integrated Au electrodes	[46]
Advantages	3×3 array of LEDs integrated on a polyimide ribbon cable and assembled with optical fibers using micromachined Si housing	[61]
	• Wireless capability: low power consumption, small size • Illumination stability • Fast switching time	
Disadvantages	• Light power issue: low coupling efficiency • Thermal and electrical leakage (direct contact with μLED) • Material issue for unverified polymer waveguide	

more adequate for therapeutic usage, as these types of devices are advantageous in securing the reliability and biocompatibility due to less or no exposure to electrically and thermally induced risks. In terms of the materials used in the devices, polymers have not yet been fully tested for their biosafety and long-term reliable operation. In this sense, utilizing commercially available optical fiber can be safer compared to polymer waveguides. However, utilizing optical fiber involves unavoidable manual and labor-intensive procedures. Multiwavelength illumination is an essential function to achieve complete control of the nerve system. However, a novel approach is required to overcome existing constraints in light delivery and device fabrication. As the optical neural interface becomes more elaborate and additional functions are integrated, more study on the stability and safety of the device will be needed. In addition, new approaches that can guarantee a simple fabrication process with high yield and throughput should also be developed.

References

1. Mansor MA, Ahmad MR (2015) Single cell electrical characterization techniques. Int J Mol Sci 16(6):12686–12712
2. Sciberras-Lim ET, Lambert AJ (2017) Attentional orienting and dorsal visual stream decline: review of behavioral and EEG studies. Front Aging Neurosci 9:246
3. Beres AM (2017) Time is of the essence: a review of electroencephalography (EEG) and event-related brain potentials (ERPs) in language research. Appl Psychophysiol Biofeedback 42:247
4. Esmaeilpour Z, Schestatsky P, Bikson M, Brunoni AR, Pellegrinelli A, Piovesan FX, Santos MM, Menezes RB, Fregni F (2017) Notes on human trials of transcranial direct current stimulation between 1960 and 1998. Front Hum Neurosci 11:71
5. Philip NS, Nelson BG, Frohlich F, Lim KO, Widge AS, Carpenter LL (2017) Low-intensity transcranial current stimulation in psychiatry. Am J Psychiatry 174(7):628–639

6. Spira ME, Hai A (2013) Multi-electrode array technologies for neuroscience and cardiology. Nat Nanotechnol 8(2):83–94
7. Lehev G, Nicolelis M (2008) State-of-the-art microwire array design for chronic neural recordings in behaving animals–methods for neural ensemble recordings. In: Nicolelis MAL (ed) Methods for neural ensemble recordings, Frontiers in neuroscience, 2nd edn. CRC Press, Boca Raton
8. Meidahl AC, Tinkhauser G, Herz DM, Cagnan H, Debarros J, Brown P (2017) Adaptive deep brain stimulation for movement disorders: the long road to clinical therapy. Mov Disord 32(6):810–819
9. Roy HA, Green AL, Aziz TZ (2018) State of the art: novel applications for deep brain stimulation. Neuromodulation 21(2):126–134
10. Semework M (2015) Microstimulation: principles, techniques, and approaches to somatosensory neuroprosthesis. Crit Rev Biomed Eng 43(1):61–95
11. Stefani A, Trendafilov V, Liguori C, Fedele E, Galati S (2017) Subthalamic nucleus deep brain stimulation on motor-symptoms of Parkinson's disease: focus on neurochemistry. Prog Neurobiol 151:157–174
12. Yu H, Neimat JS (2008) The treatment of movement disorders by deep brain stimulation. Neurotherapeutics 5(1):26–36
13. Hajioff D (2016) Cochlear implantation: a review of current clinical practice. Br J Hosp Med (Lond) 77(12):680–684
14. Cosetti MK, Waltzman SB (2012) Outcomes in cochlear implantation: variables affecting performance in adults and children. Otolaryngol Clin North Am 45(1):155–171
15. Luo YH, da Cruz L (2016) The Argus((R)) II retinal prosthesis system. Prog Retin Eye Res 50:89–107
16. Chader GJ, Weiland J, Humayun MS (2009) Artificial vision: needs, functioning, and testing of a retinal electronic prosthesis. Prog Brain Res 175:317–332
17. Rothschild G, Nelken I, Mizrahi A (2010) Functional organization and population dynamics in the mouse primary auditory cortex. Nat Neurosci 13:353–360
18. Roberts TF, Tschida KA, Klein ME, Mooney R (2010) Rapid spine stabilization and synaptic enhancement at the onset of behavioural learning. Nature 463(7283):948–952
19. Clancy KB, Koralek AC, Costa RM, Feldman DE, Carmena JM (2014) Volitional modulation of optically recorded calcium signals during neuroprosthetic learning. Nat Neurosci 17(6):807–809
20. Flusberg BA, Jung JC, Cocker ED, Anderson EP, Schnitzer MJ (2005) In vivo brain imaging using a portable 3.9 gram two-photon fluorescence microendoscope. Opt Lett 30(17):2272–2274
21. Sawinski J, Denk W (2007) Miniature random-access fiber scanner for in vivo multiphoton imaging. J Appl Phys 102(3):034701
22. Engelbrecht CJ, Johnston RS, Seibel EJ, Helmchen F (2008) Ultra-compact fiber-optic two-photon microscope for functional fluorescence imaging in vivo. Opt Express 16(8):5556–5564
23. Piyawattanametha W, Cocker ED, Burns LD, Barretto RPJ, Jung JC, Ra H, Solgaard O, Schnitzer MJ (2009) In vivo brain imaging using a portable 2.9 g two-photon microscope based on a microelectromechanical systems scanning mirror. Opt Lett 34(15):2309–2311
24. Levene MJ, Dombeck DA, Kasischke KA, Molloy RP, Webb WW (2004) In vivo multiphoton microscopy of deep brain tissue. J Neurophysiol 91(4):1908–1912
25. Jung JC, Mehta AD, Aksay E, Stepnoski R, Schnitzer MJ (2004) In vivo mammalian brain imaging using one- and two-photon fluorescence microendoscopy. J Neurophysiol 92(5):3121–3133
26. Llewellyn ME, Barretto RPJ, Delp SL, Schnitzer MJ (2008) Minimally invasive high-speed imaging of sarcomere contractile dynamics in mice and humans. Nature 454(7205):784–788
27. Bocarsly ME, Jiang W-C, Wang C, Dudman JT, Ji N, Aponte Y (2015) Minimally invasive microendoscopy system for in vivo functional imaging of deep nuclei in the mouse brain. Biomed Opt Express 6(11):4546–4556

28. Deisseroth K (2015) Optogenetics: 10 years of microbial opsins in neuroscience. Nat Neurosci 18(9):1213–1225
29. Warden MR, Cardin JA, Deisseroth K (2014) Optical neural interfaces. Annu Rev Biomed Eng 16:103–129
30. Adamantidis AR, Zhang F, Aravanis AM, Deisseroth K, De Lecea L (2007) Neural substrates of awakening probed with optogenetic control of hypocretin neurons. Nature 450(7168):420–429
31. Aravanis AM, Wang LP, Zhang F, Meltzer LA, Mogri MZ, Schneider MB, Deisseroth K (2007) An optical neural interface: in vivo control of rodent motor cortex with integrated fiberoptic and optogenetic technology. J Neural Eng 4(3):S143–S156
32. Stark E, Koos T, Buzsaki G (2012) Diode probes for spatiotemporal optical control of multiple neurons in freely moving animals. J Neurophysiol 108(1):349–363
33. Chen SY, Pei WH, Gui Q, Chen YF, Zhao SS, Wang H, Chen HD (2013) A fiber-based implantable multi-optrode array with contiguous optical and electrical sites. J Neural Eng 10(4):046020
34. Wu F, Stark E, Im M, Cho IJ, Yoon ES, Buzsaki G, Wise KD, Yoon E (2013) An implantable neural probe with monolithically integrated dielectric waveguide and recording electrodes for optogenetics applications. J Neural Eng 10(5):056012
35. Rubehn B, Wolff SBE, Tovote P, Luthi A, Stieglitz T (2013) A polymer-based neural microimplant for optogenetic applications: design and first in vivo study. Lab Chip 13(4):579–588
36. Abaya TV, Blair S, Tathireddy P, Rieth L, Solzbacher F (2012) A 3D glass optrode array for optical neural stimulation. Biomed Opt Express 3(12):3087–3104
37. LeChasseur Y, Dufour S, Lavertu G, Bories C, Deschenes M, Vallee R, De Koninck Y (2011) A microprobe for parallel optical and electrical recordings from single neurons in vivo. Nat Methods 8(4):319–325
38. Anikeeva P, Andalman AS, Witten I, Warden M, Goshen I, Grosenick L, Gunaydin LA, Frank LM, Deisseroth K (2012) Optetrode: a multichannel readout for optogenetic control in freely moving mice. Nat Neurosci 15(1):163–170
39. Wang J, Wagner F, Borton DA, Zhang J, Ozden I, Burwell RD, Nurmikko AV, van Wagenen R, Diester I, Deisseroth K (2012) Integrated device for combined optical neuromodulation and electrical recording for chronic in vivo applications. J Neural Eng 9(1):016001
40. Son Y, Lee HJ, Kim J, Lee CJ, Yoon E-S, Kim TG, Cho I-J (2015) A new monolithically integrated multi-functional MEMS neural probe for optical stimulation and drug delivery. Presented at the 28th IEEE international conference on micro electro mechanical systems (MEMS), Estoril, 18–22 January 2015
41. Zorzos AN, Scholvin J, Boyden ES, Fonstad CG (2012) Three-dimensional multiwaveguide probe array for light delivery to distributed brain circuits. Opt Lett 37(23):4841–4843
42. Kim TI, McCall JG, Jung YH, Huang X, Siuda ER, Li YH, Song JZ, Song YM, Pao HA, Kim RH, Lu CF, Lee SD, Song IS, Shin G, Al-Hasani R, Kim S, Tan MP, Huang YG, Omenetto FG, Rogers JA, Bruchas MR (2013) Injectable cellular-scale optoelectronics with applications for wireless optogenetics. Science 340(6129):211–216
43. Cao H, Gu L, Mohanty SK, Chiao JC (2013) An integrated mu LED optrode for optogenetic stimulation and electrical recording. IEEE Trans Biomed Eng 60(1):225–229
44. Kwon KY, Sirowatka B, Weber A, Li W (2013) Opto-mu ECoG array: a hybrid neural Interface with transparent mu ECoG electrode array and integrated LEDs for optogenetics. IEEE Trans Biomed Circuits Syst 7(5):593–600
45. Fan B, Kwon KY, Weber A, Li W (2014) An implantable, miniaturized SU-8 optical probe for optogenetics-based deep brain stimulation. Presented at the 36th international conference of the IEEE engineering in medicine and biology society, Chicago, 26–30 August 2014
46. Kwon KY, Lee HM, Ghovanloo M, Weber A, Li W (2015) Design, fabrication, and packaging of an integrated, wirelessly-powered optrode array for optogenetics application. Front Syst Neurosci 9:69

47. Fan B, Kwon KY, Rechenberg R, Becker MF, Weber AJ, Li W (2016) A hybrid neural interface optrode with a polycrystalline diamond heat spreader for optogenetics. Technology 4(1):15–22
48. Fan B, Li W (2015) Miniaturized optogenetic neural implants: a review. Lab Chip 15(19):3838–3855
49. McCall JG, Kim TI, Shin G, Huang X, Jung YH, Al-Hasani R, Omenetto FG, Bruchas MR, Rogers JA (2013) Fabrication and application of flexible, multimodal light-emitting devices for wireless optogenetics. Nat Protoc 8(12):2413–2428
50. Lee ST, Williams PA, Braine CE, Lin DT, John SWM, Irazoqui PP (2015) A miniature, fiber-coupled, wireless, deep-brain optogenetic stimulator. IEEE Trans Neural Syst Rehabil Eng 23(4):655–664
51. Iwai Y, Honda S, Ozeki H, Hashimoto M, Hirase H (2011) A simple head-mountable LED device for chronic stimulation of optogenetic molecules in freely moving mice. Neurosci Res 70(1):124–127
52. Wentz CT, Bernstein JG, Monahan P, Guerra A, Rodriguez A, Boyden ES (2011) A wirelessly powered and controlled device for optical neural control of freely-behaving animals. J Neural Eng 8(4):046021
53. Dupré L, Marra M, Verney V, Aventurier B, Henry F, Olivier F, Tirano S, Daami A, Templier F (2017) Processing and characterization of high resolution GaN/InGaN LED arrays at 10 micron pitch for micro display applications. In: Proceedings of the SPIE, pp 1010422–1010421
54. Boyden ES, Zhang F, Bamberg E, Nagel G, Deisseroth K (2005) Millisecond-timescale, genetically targeted optical control of neural activity. Nat Neurosci 8(9):1263–1268
55. Zhang F, Wang LP, Brauner M, Liewald JF, Kay K, Watzke N, Wood PG, Bamberg E, Nagel G, Gottschalk A, Deisseroth K (2007) Multimodal fast optical interrogation of neural circuitry. Nature 446(7136):633–634
56. Gradinaru V, Zhang F, Ramakrishnan C, Mattis J, Prakash R, Diester I, Goshen I, Thompson KR, Deisseroth K (2010) Molecular and cellular approaches for diversifying and extending optogenetics. Cell 141(1):154–165
57. Yizhar O, Fenno LE, Davidson TJ, Mogri M, Deisseroth K (2011) Optogenetics in neural systems. Neuron 71(1):9–34
58. Tye KM, Mirzabekov JJ, Warden MR, Ferenczi EA, Tsai HC, Finkelstein J, Kim SY, Adhikari A, Thompson KR, Andalman AS, Gunaydin LA, Witten IB, Deisseroth K (2013) Dopamine neurons modulate neural encoding and expression of depression-related behaviour. Nature 493(7433):537–541
59. Bernstein JG, Allen BD, Guerra AA, Boyden ES (2015) Processes for design, construction and utilisation of arrays of light-emitting diodes and light-emitting diode-coupled optical fibres for multi-site brain light delivery. J Eng http://dx.doi.org/10.1049/joe.2014.0304
60. Schwaerzle M, Elmlinger P, Paul O, Ruther P (2015) Miniaturized 3×3 optical fiber array for optogenetics with integrated 460 nm light sources and flexible electrical interconnection. Presented at the 28th IEEE international conference on micro electro mechanical systems (MEMS), Estoril, 18–22 January 2015
61. Hudson MC (1974) Calculation of maximum optical coupling efficiency into multimode optical-waveguides. Appl Opt 13(5):1029–1033
62. Zhang JY, Laiwalla F, Kim JA, Urabe H, Van Wagenen R, Song YK, Connors BW, Zhang F, Deisseroth K, Nurmikko AV (2009) Integrated device for optical stimulation and spatiotemporal electrical recording of neural activity in light-sensitized brain tissue. J Neural Eng 6(5):055007
63. Royer S, Zemelman BV, Barbic M, Losonczy A, Buzsaki G, Magee JC (2010) Multi-array silicon probes with integrated optical fibers: light-assisted perturbation and recording of local neural circuits in the behaving animal. Eur J Neurosci 31(12):2279–2291
64. Voigts J, Siegle JH, Pritchett DL, Moore CI (2013) The flexDrive: an ultra-light implant for optical control and highly parallel chronic recording of neuronal ensembles in freely moving mice. Front Syst Neurosci 7:8
65. Andersen P, Moser EI (1995) Brain temperature and hippocampal function. Hippocampus 5(6):491–498

66. Fekete Z, Csernai M, Kocsis K, Horvath AC, Pongracz A, Bartho P (2017) Simultaneous in vivo recording of local brain temperature and electrophysiological signals with a novel neural probe. J Neural Eng 14(3):034001
67. McAlinden N, Massoubre D, Richardson E, Gu E, Sakata S, Dawson MD, Mathieson K (2013) Thermal and optical characterization of micro-LED probes for in vivo optogenetic neural stimulation. Opt Lett 38(6):992–994

Scalable Nanofabrication of Plasmonic Nanostructures for Trace-Amount Molecular Sensing Based on Surface-Enhanced Raman Spectroscopy (SERS)

Seunghee H. Cho, Kwang Min Baek, and Yeon Sik Jung

1 Introduction

Detection of trace-amount molecules has become an important issue in various fields such as environmental pollution [1], security and counterterrorism [2], food safety [3], and human healthcare [4–6]. Harmful chemicals have notorious effects even at trace-amount concentrations due to their bioaccumulation over long periods of time [7–11]. Traditional and still widely used analysis tools for low-concentration molecules in the environment include gas chromatography [12] and mass spectrometry [13–15]. Gas chromatography collects information of their physical and chemical properties, and mass spectrometry derives elemental information. Although useful for accurate analysis, most of these technologies require complex and destructive methods for sample preparation. In biomedical fields, trace-amount sensing is also crucial in the early diagnosis of diseases through the detection of biomarkers [16]. Current techniques mostly utilize antibiotic and enzymatic reactions using cell-culture methods [17]. Yet, the determination of biological species within a sample is limited to using blood-drawn samples, which requires a facile and noninvasive method of sensing. Furthermore, the costly equipment and the time-consuming nature of measurements make it difficult to access the technology. Hence, next-generation methods for universal molecular detection and analysis are in great demand to be applied in various fields.

Meanwhile, Raman spectroscopy is a widely used analysis tool first introduced by Sir C. V. Raman to identify molecular structures by detecting the Raman scattering phenomenon [18, 19], a form of inelastic scattering when incident photons hit a molecule. An extremely small portion ($\sim 10^{-7}$) of the incident photons

S. H. Cho · K. M. Baek · Y. S. Jung (✉)
Department of Materials Science and Engineering, Korea Advanced Institute of Science and Technology (KAIST), Daejeon, Republic of Korea
e-mail: ysjung@kaist.ac.kr

© Springer Nature Switzerland AG 2020
Y. Liu et al. (eds.), *Smart Sensors and Systems*,
https://doi.org/10.1007/978-3-030-42234-9_4

scatters inelastically by interacting with the vibrational and rotational states, reading the molecular structure of the sample. The resulting Raman spectrum, therefore, serves as a characteristic fingerprint of a specific molecule, allowing highly accurate identification of unknown molecules [20, 21]. Also, Raman spectroscopy is highly advantageous in rapid and nondestructive analysis for a wide range of materials in solid, liquid, and gas forms. However, Raman signals are weak in intensity due to the small probability of Raman scattering and thus lack low-concentration sensitivity [22]. In order to resolve this issue, extensive research has been focused on enhancing the Raman signal intensity. In 1977, it was reported that noble metals can strengthen Raman signals substantially by amplifying the local electromagnetic field through plasmonic effects [23]. In addition, it has been demonstrated that the morphologies of metallic structures play a key role in generating localized surface plasmon resonance (LSPR) effects for further enhancement of signal intensities [24–34]. This technique, where the surface morphology of metallic structures is modified to increase the Raman signal intensity by several orders of magnitude, has been named surface-enhanced Raman spectroscopy (SERS) (Fig. 1). Later studies on plasmonic nanostructures for SERS have shown single-molecule-level sensitivities [35, 36], and thus SERS has risen as a strong candidate for the detection of trace-amount molecules owing to its rapid, facile, nondestructive nature of measurement as well as its high sensitivity [37–41].

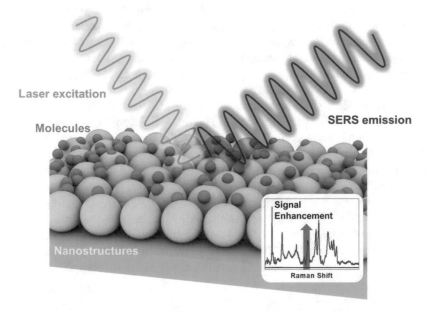

Fig. 1 Schematic illustration of surface-enhanced Raman spectroscopy (SERS) analysis

The main purpose of this chapter is to provide a description of various plasmonic nanostructures for SERS and their state-of-the-art fabrication techniques. First we briefly explain the mechanism of SERS and the important factors that contribute to signal improvement, and we then describe various approaches for fabricating plasmonic nanostructures: top–down, bottom–up, and a combination of both.

2 SERS Enhancement Mechanism

Signal enhancement in SERS mainly depends on the LSPR effect. LSPR, a non-propagating surface plasmon, occurs at the surface of an isolated metallic nanostructure [42]. For example, at the curved surface of a metal nanoparticle, the electron cloud oscillates upon incident light due to the displacement from the electromagnetic wave and restoring force from the coulombic attraction with the nuclei [30, 43]. Collective electron oscillation along the metallic surface forms "hot spots" [44] where local electromagnetic fields are greatly amplified (Fig. 2a). To observe the electric field enhancements close to hot spots, calculation methods

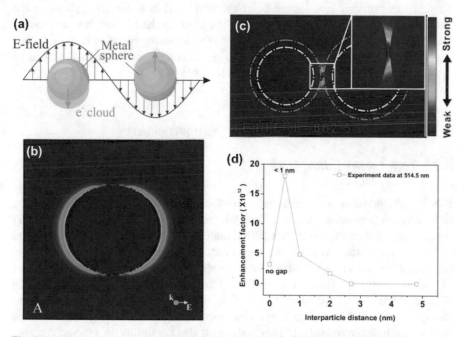

Fig. 2 Mechanism of localized surface plasmon resonance (LSPR) and electric field enhancement. (a) Plasmon oscillation between metal spheres. (b) Simulation of electric field near a metal sphere. (Reprinted with permission from Ref. [43]. Copyright (2003), ACS). (c) Electromagnetic field simulation of nanogap between particles. (d) Enhancement factor changes by interparticle distance. (Reprinted with permission from Ref. [45]. Copyright (2012), ACS)

such as finite-difference time–domain (FDTD) can be used. Using different shades of contour, the distribution of the electric field nearby a metallic nanoparticle is illustrated in Fig. 2b. Also, electric field enhancement may be further increased by forming nano-sized gaps between metallic nanostructures. Studies have tested the enhancement of SERS signals for nanoparticles with different nanogap sizes [45]. In particular, FDTD calculations show maximum electric field strength at the smallest gap between two nanoparticles (Fig. 2c). Experimental data also confirmed the strongest enhancement for gaps smaller than 1 nm compared to structures with no gap and nanogaps larger than 1 nm (Fig. 2d). Though LSPR occurs at the surface, LSPR is mainly generated in the near-field, and thus electromagnetic fields are most enhanced within a nanoscale range of the surface. In addition, the strong enhancement is partially attributed to the resonance properties due to the fact that the electromagnetic field enhancement is maximal when the resonance frequency matches that of the incident photons [46–49]. The resonance frequency of a metallic nanostructure is determined by the size and shape, and thus careful and systematic tuning of morphology can contribute to additional enhancements.

For the design of effective SERS structures, there are various key principles. Most importantly, plasmonic nanostructures should demonstrate strong enhancement of electromagnetic fields. Also, the reproducibility and stability are also crucial in Raman spectroscopy measurements of chemicals. This is directly related to the second principle, use of an efficient fabrication method. Complex and costly fabrication methods have often been an issue for designing SERS structures. In this regard, ideal SERS structures must have high enhancement factors and should be fabricated with a cost-effective, facile, and scalable process.

3   Nanofabrication of Plasmonic Nanostructures

3.1   Top–Down Lithography

E-beam lithography is a powerful and extensively used technique for the fabrication of high-resolution nanostructures [50, 51]. Its advantages include the ability to freely design desired nanostructures with high reliability. Such advantages have been used to explore the fundamental physics behind the amplification of the electric field at nanoscale hot spots near metallic nanostructures [52]. Au plasmonic nanostructures of various morphologies with sub-10 nm gaps have been fabricated, as shown in Fig. 3a–d. Arrays of Au squares, triangles, bowties, and trimers with controlled nanogaps were designed (Fig. 3a–d), and as the nanogap size decreased, resonance bands significantly redshifted. This confirmed the possibility of precisely tuning the plasmon resonance. Among Au nanohexagons, nanosquares, and nanotriangles that were fabricated with optimized nanogap sizes, nanotriangles have shown the largest SERS enhancement. Raman measurement results of 4-aminothiophenol (4-ATP) using various morphologies confirm the effect of nanoparticle shape, where

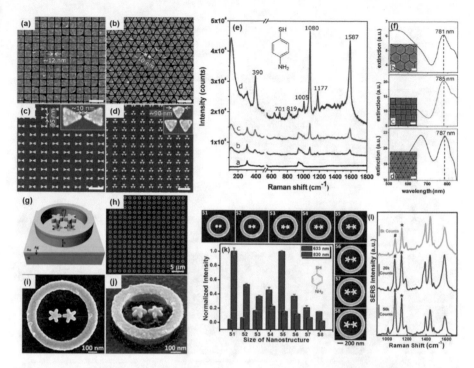

Fig. 3 (**a–d**) SEM images of various Au nanostructures: (**a**) nanosquares, (**b**) nanotriangles, (**c**) bowties, and (**d**) trimers. (**e, f**) SERS spectra of 4-ATP. (Reprinted with permission from Ref. [52]. Copyright (2011), ACS.) (**g–j**) Schematic and SEM images of nanostar dimers. (**k, l**) SERS intensity of p-aminothiophenol (pATP) of different sizes and excitation wavelengths. (Reprinted with permission from Ref. [53]. Copyright (2014), ACS)

the sharp edges in triangles had the strongest enhancement [54] (Fig. 3e). All nanostructures were designed to have plasmon resonance in a spectral alignment with the excitation laser wavelength of 785 nm (Fig. 3f). Three-dimensional structures can also be fabricated via e-beam lithography. Three-dimensional nanostar dimers of different sizes were fabricated to observe the effect on plasmon resonance properties [53]. Bimetallic AgAu nanostar dimers were lithographically produced in a ring, as shown in Fig. 3g–j. SERS measurements with excitation wavelengths of 532, 633, and 830 nm have shown maximum enhancement at different sizes of nanostars (Fig. 3k–l). This demonstrates the tunability of plasmon resonance by controlling the nanostructure dimensions to match the excitation laser.

Nanoimprint lithography is a useful method to fabricate highly reproducible nanostructures. As shown in SEM images, gold fingers with a high aspect ratio have been fabricated to utilize the capillary force upon dropping an analyte solution and trigger coalescence of the nanofingers [55] (Fig. 4a–d). The self-closing behavior of gold fingers is capable of trapping trans-1,2-bis(4-pyridyl)ethylene (BPE) molecules between the gold tips for SERS enhancement, as shown in Fig. 4e. To clarify the effect of entrapment, peak shifting in Raman spectra of BPE for preclosed and

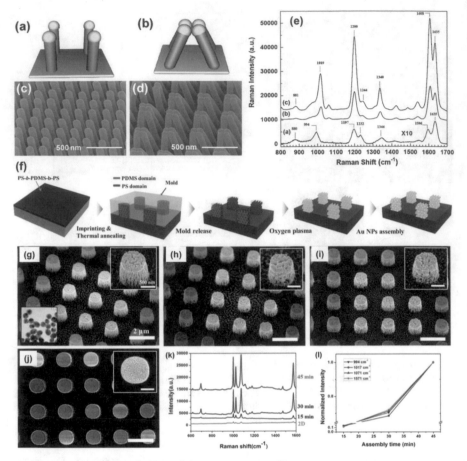

Fig. 4 (a–d) Schematic and SEM images of self-closing Au nanofingers. (e) SERS spectra of preclosed and BPA-trapped nanofingers. (Reprinted with permission from Ref. [55]. Copyright (2011), ACS.) (f) Schematic illustration of fabrication process for Au-decorated nanoporous SiO$_2$ microcylinders. (g–i) SEM image of Au-decorated nanoporous SiO$_2$ microcylinders of different deposition times: (g) 15, (h) 30, (i) 45 min. (j) Dense microcylinders decorated with Au. (k, l) SERS spectra of BT with Au-decorated nanoporous SiO$_2$ microcylinders of different deposition times. (Reprinted with permission from Ref. [56]. Copyright (2013), ACS)

BPE-trapped fingers was observed. Compared to preclosed fingers, upright fingers that have undergone capillary–force-driven solution drying showed significantly stronger Raman signals. This confirmed that BPE molecules must be placed at the sub-nanometer gaps between gold tips and also the bonding of BPE to gold tips should be formed. Porous plasmonic nanostructures can also be fabricated from imprint-based lithography. An array of nanoporous SiO$_2$ microcylinders has been produced by micro-imprinting block copolymers and decoration of Au nanoparticles to form SERS-active structures [56] (Fig. 4f). Nanoporous structures improve the diffusion of analyte solutions into nanoscale hot spots of a SERS structure and thus are advantageous for Raman measurements. SEM images in Fig. 4g–j clearly show how Au particles penetrate the porous templates and also that thickness and density are increased with longer Au deposition times. Densely packed Au nanoparticles generate plasmonic coupling due to LSPR, greatly enhancing the Raman signals of benzenethiol (BT) molecules (Fig. 4k–l).

Even without optical lithographic techniques, plasmonic nanostructures can be fabricated by applying surface treatments to various substrates. Ag-coated Si nanowires were fabricated using close-packed nanosphere arrays as a mask for metal-assisted chemical etching [57]. Figure 5a shows the morphology of hexagonally packed nanowires and the Ag–shell–Si–core structures. Despite the relatively large size (110 nm in diameter, 700 nm in length), Ag-coated Si nanowires are an example of a nanogap-free SERS system. Unlike non-propagating LSPR of dipolar plasmon resonance [59], continuous Ag shells create propagating surface plasmons with a higher mode of plasmon excitation [60]. FDTD calculations of

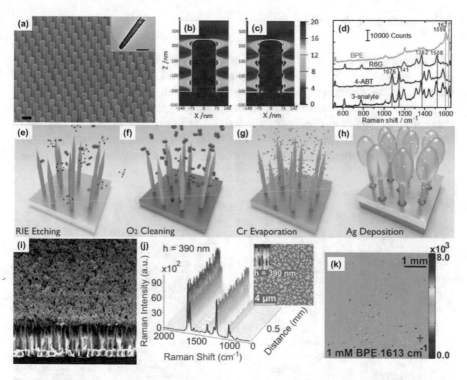

Fig. 5 (**a**) SEM image of Ag-coated Si nanowires. Inset is TEM image. (**b, c**) FDTD simulation of electric field for smooth and rough Ag-coated Si nanowires. (**d**) SERS spectra of BPE, 4-ABT, and R6G. (Reprinted with permission from Ref. [57]. Copyright (2013), ACS.) (**e–h**) Fabrication process of Ag-capped Si nanopillars. (**i**) SEM image of Ag-capped Si nanopillars. (**j**) SERS spectra of BPE. Inset is uniform distribution (side view) and aggregation of nanopillars (top view) before and after detection. (**k**) SERS signal uniformity across large area. (Reprinted with permission from Ref. [58]. Copyright (2015), ACS)

electric fields in Fig. 5b, c exhibit six antinodes on the surface of Ag-coated Si nanowires for both smooth and rough structures (Fig. 5b, c). As demonstrated in contoured shades, surface roughness contributed to further enhancement of electric fields. Raman measurements using three different analytes each show significant SERS enhancement up to an enhancement factor of $\sim 10^6$ (Fig. 5d). Wafer-scale SERS structures can also be fabricated by taking advantage of non-lithographic surface treatment methods. High-aspect-ratio Si pillar arrays were easily fabricated by maskless reactive ion etching (RIE), and subsequent Ag deposition formed Ag capping layers for SERS analysis [58] (Fig. 5e–i). According to finite element method (FEM) calculations, plasmon coupling in the Ag cap cavity and Ag–Si material interface are the largest factors for SERS enhancement (Fig. 5j). In addition, the maskless etching process allows large-scale fabrication and also shows reliable uniformity across a 5×5 mm^2 substrate.

3.2 Bottom–Up Synthesis

The simple solution-phase synthesis of metal nanoparticles is beneficial in fabricating plasmonic SERS structures. It provides control over the shape and size of the particles by changing synthesis parameters, which also enables fine-tuning of the plasmon resonance wavelength. As branches are grown from Au nanoparticles, local electric fields are significantly enhanced at the tips of branches, leading to highly intensified Raman signals [61] (Fig. 6a–j). Silica-coated Au nanoparticles of different shapes (spheres, rods, and triangles) were used as templates to grow branch

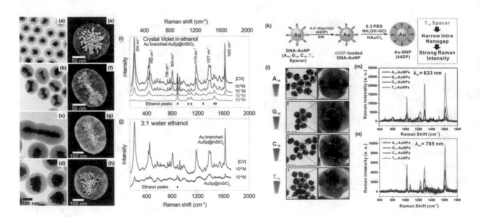

Fig. 6 (**a–d**) 2D and (**e–h**) 3D TEM characterization of Au nanoparticles with different morphologies. (**i, j**) SERS spectra of CV before and after templated branching of Au nanoparticles. (Reprinted with permission from Ref. [61]. Copyright (2015), ACS.) (**k**) Schematic illustration of Au nanoparticle synthesis with intra-nanogaps. (**l**) Solution color and TEM images. (**m, n**) SERS spectra of excitation wavelengths 633 and 785 nm. (Reprinted with permission from Ref. [62]. Copyright (2015), ACS)

tips using a solution-phase method. This improved the surface-to-volume ratio, which increased analyte adsorption and produced higher enhancement of electric fields by similar mechanisms as demonstrated for Au nanostars in previous research [63]. As shown in Fig. 6i, j, from Raman measurements of crystal violet (CV), the detection limit of branched Au nanoparticles was 10^{-6} M, and signal intensities were increased by one order of magnitude compared to spherical particles. Other than changing the shape of metal nanoparticles, narrow nanogaps can be formed within the particle for a coupling effect. Au nanoparticles with intra-nanogaps have also been reported to successfully tune the plasmon absorption for SERS enhancement at 785 nm excitation [62]. During their synthesis, Au nanoparticles are surface-modified with DNA containing a spacer sequence, and the outer Au shell is grown with solution-phase methods, as shown in Fig. 6k. SERS enhancement is derived from the strong plasmon coupling between the Au core and shell, and thus narrower gaps produce larger enhancement. As shown in Fig. 6, transmission electron microscopy (TEM) images clearly demonstrate the presence of nanogaps among Au nanoparticles synthesized using various spacer sequences. Among those, Au nanoparticles prepared using thymine spacers produced Raman signal intensities more than seven times higher than the others (Fig. 6n). This can be supported by the sub-nanometer gaps formed with thymine, unlike other spacers, as presented in the TEM images and also by plasmon resonance that better matches the 785 nm excitation.

Although isolated metallic nanoparticles can generate plasmonic resonance, the effects can be further enhanced by increasing the number and areal density of hot spots. In this aspect, self-assembly of metallic particles is a simple but effective technique to achieve drastically improved SERS enhancements. A vertical monolayer of Au nanorods has been fabricated by a simple and robust evaporation-induced self-assembly strategy [64]. As shown in Fig. 7a, b, the plasmon band of vertically standing close-packed Au nanorods is located near 717 nm and SEM images confirm monolayer formation. Nanogap sizes were reduced by changing the ionic strength of the Au nanorod solution, which is expressed in Debye length, and further UV ozone treatments created sub-nanometer gaps by removing surfactant molecules to bring the nanoparticles closer together (Fig. 7c–e). Raman measurements with a vertical monolayer of Au nanorods were used to successfully detect trace amounts of the food contaminants benzyl butyl phthalate (BBP, $C_{19}H_{20}O_4$) and bis(2-ethylhexyl)phthalate (DEHP, $C_{24}H_{38}O_4$) in orange juice (Fig. 7g). Metal nanoparticles can also be self-assembled with spacer molecules to control and rigidly fix the nano-sized junctions [65]. Au nanoparticle clusters are fabricated using a hollow macrocyclic spacer known as cucurbit[n]uril (CB[n], $n = 5$–8, 10) to form precisely sized nanojunctions of 0.9 nm (Fig. 7h). Raman measurements were conducted to detect the photochemical activity and inclusion of diaminostilbene (DAS) within Au nanoparticle–CB[n] clusters upon UV irradiation. Figure 7i, j shows SERS spectra of DAS inserted into the nanojunctions of Au nanoparticle clusters with increasing UV irradiation time. The Au nanoparticle clusters demonstrate large SERS enhancements due to the hot spots formed at the nanogaps and also the plasmon resonance modes across the chain [66].

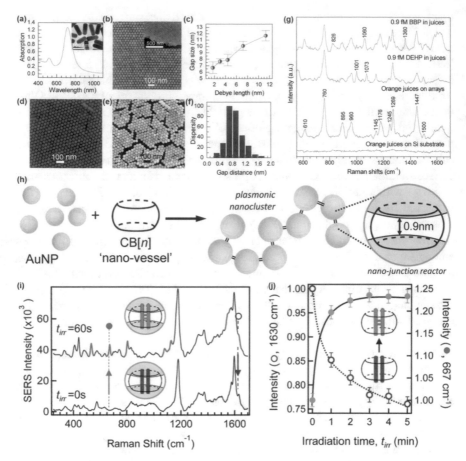

Fig. 7 (**a**) Absorption spectrum of Au nanorods. Inset is TEM image. (**b**) SEM image of hexagonally arranged vertical Au nanorod monolayer. (**c**) Edge-to-edge gap distance plotted by Debye length. (**d, e**) SEM image of hexagonally arranged vertical Au nanorod monolayers: (**d**) before and (**e**) after UV treatment. (**f**) Nanogap size distribution after UV treatment. (**g**) SERS spectra of orange juice containing BBP and DEHP. (Reprinted with permission from Ref. [64]. Copyright (2013), ACS.) (**h**) Schematic illustration of self-assembled Au nanoparticle nanoclusters with fixed separation. (**i, j**) SERS spectra of DAS inserted into Au nanoparticle nanoclusters after UV irradiation and intensities at 1630 cm^{-1} by UV irradiation time. (Reprinted with permission from Ref. [65]. Copyright (2013), ACS)

Colloidal or nanosphere lithography is a scalable and cost-effective tool for reproducible fabrication of submicron structures with dense plasmonic nanogaps. A low-cost fabrication strategy for annular cavity arrays has been reported using polystyrene (PS) spheres [67]. After embedding hexagonally packed PS spheres in silica gel, PS spheres were partially etched using reactive ion etching (RIE), and Ag deposition was carried out to form annular cavity arrays (Fig. 8). Along with electric field enhancement from nanogaps in the cavities, cylindrical surface plasmon

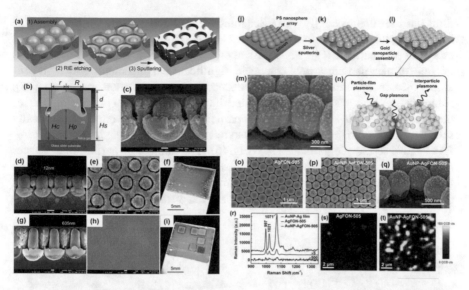

Fig. 8 (**a**) Schematic illustration of fabrication of annular cavity arrays. (**b, c**) Cross-section model and SEM image of annular cavity arrays. (**d–i**) Annular cavity arrays fabricated with different dimensions: (**d, g**) SEM cross sections, (**e, h**) SEM top views, (**f, i**) photographs. (Reprinted with permission from Ref. [67]. Copyright (2015), ACS.) (**j–l**) Fabrication process of Au nanoparticle-decorated Ag film over nanosphere structure. (**m**) SEM image. (**n**) Illustration of SERS effect. (**o–q**) SEM image top views and cross-view of Au nanoparticle-decorated Ag film over nanospheres: (**o**) before and (**p**) after Au decoration. (**r**) SERS spectra of Au nanoparticles on Ag film, Ag film on nanosphere, and Au nanoparticle-decorated Ag film on nanosphere structure. (**s, t**) Raman mapping of Au nanoparticle-decorated Ag film over nanosphere before and after Au decoration. (Reprinted with permission from Ref. [68]. Copyright (2015), ACS)

resonance properties [69] of annular cavity array structures have attracted strong interest for SERS. The tunability of geometry parameters such as nanogap width, depth, and radius was demonstrated to modulate the plasmon resonance across the entire optical wavelength range (Fig. 8d–i). Hybrid plasmonic nanostructures can also be fabricated using nanosphere lithography. As shown in Fig. 8j–q, Au nanoparticle-decorated Ag films over nanosphere structures have been produced using hexagonally close-packed PS spheres, Ag deposition, and Au nanoparticle decoration [68]. SERS spectra of the fabricated structures using BT molecules showed significant enhancements compared to those without Au nanoparticles. The increased Raman signals can be attributed to not only the particle-film plasmon coupling but also the interparticle coupling generated by the hybrid structure.

Block copolymer self-assembly is a simple method to produce large-area, ordered nanostructures of various morphologies [70–72]. By incorporating metallic elements, nanostructures with highly tunable plasmonic resonance can be achieved. Hexagonal arrays of Au–Ag core–shell and alloy nanoparticles have been fabricated using block copolymer self-assembly [73]. Solution-phase Au and Ag incorporation into self-assembled polystyrene-block-poly(4-vinylpyridine) (PS-*b*-P4VP) thin films has been conducted (Fig. 9a). Scanning electron microscopy (SEM) and TEM images present successful formation of Au and Au–Ag nanostructures in hexagonal arrays of nanopatterns (Fig. 9b–g). As shown in Fig. 9h, i, Raman measurements of Au–Ag nanoparticle arrays displayed significant enhancements and high reliability due to systematically tuned broadband surface plasmon resonance [75]. Nanoparticles can also be self-assembled into arrays of clusters with the aid of block copolymer self-assembly. Au nanoparticle cluster arrays were electrostatically self-assembled onto the surface of polystyrene-block-poly(2-vinylpyridine) (PS-*b*-P2VP) reverse micelle particles [74] (Fig. 9j). The number of particles per cluster (Fig. 9k–r) and intercluster separation distance have been controlled, where the Raman signal enhancement ratio was maximized for the highest number of particles per cluster and the smallest intercluster separation distance (Fig. 9s, t).

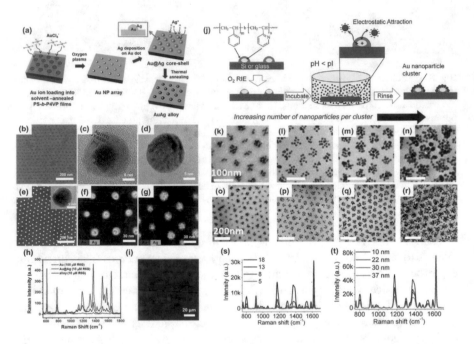

Fig. 9 (**a**) Schematic illustration of Au–Ag core–shell nanopatterning process. (**b**) SEM image of solvent-annealed PS-*b*-P4VP film. (**c, d**) TEM image of Au–Ag core–shell nanoparticles. (**e**) SEM image of Au–Ag core–shell nanoparticle arrays. Inset is TEM image. (**f, g**) EDS elemental mapping. (**h**) SERS spectra, and (**i**) Raman mapping image of R6G. (Reprinted with permission from Ref. [73]. Fabrication process of Au nanoparticle cluster. Copyright (2015), ACS.) (**j**) Fabrication process of Au nanoparticle cluster arrays. (**k–r**) TEM images of Au nanoparticle clusters with increasing number of nanoparticles per cluster. (**s, t**) SERS spectra of CV using Au nanoparticle clusters: (**s**) different numbers of nanoparticles per cluster, and (**t**) different intercluster separations. (Reprinted with permission from Ref. [74]. Copyright (2012), ACS)

3.3   Combined Approach

In recent years, a combined approach of top–down lithography and bottom–up synthesis has been reported, namely, directed self-assembly of block copolymers [76]. Using predesigned trenches, block copolymer films form self-assembled nanostructures with controlled position, orientation, and geometry [77–79]. Au nanowires and nanorods were fabricated to tune the plasmonic resonance and improve the Raman signals of molecules [80]. After self-assembled block copolymer nanowire structures are treated using plasma, polymer replicas function as a medium for Au deposition using an evaporator. By carefully controlling the deposition conditions, either continuous nanowires or discrete nanorods can be produced (Fig. 10a, e). As shown in Fig. 10b–d, e–h, average nanogap sizes below 5 nm are highly effective hot spots for SERS signal enhancement, and thus Raman signals for R6G using nanorods show the highest enhancement (Fig. 10j). Self-assembly guiding trenches may have different geometries such as hexagonal packed circular mesh patterns [81]. Hexagonally close-packed PS spheres were used as masks for Au deposition, and block copolymer self-assembly is conducted within circular trenches followed by an additional Au deposition on the polymeric cylinder nanostructures. As shown in SEM images, cylinders guided by circular trenches resulted in concentric rings, which varied in number of rings depending on the trench dimensions (Fig. 10k–n). With an increasing number of rings, the areal density of the nanogaps

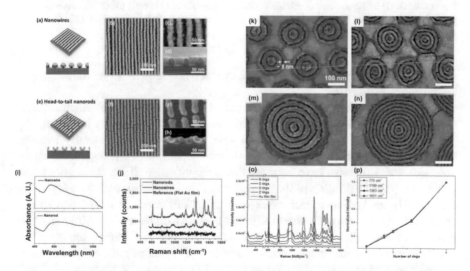

Fig. 10 (**a–h**) Schematic representation and SEM images of Au nanowires and Au nanorods. (**i**) Absorption spectra of Au nanowires and Au nanorods. (**j**) SERS spectra of R6G using Au nanowires and Au nanorods. (Reprinted with permission from Ref. [80].) (**k–n**) Controlled number of cylinders per circular trench for concentric Au nanoring structures. (**o**) SERS spectra of R6G using concentric Au nanorings. (**p**) SERS intensity with increasing number of rings. (Copyright (2016), Wiley. Reprinted with permission from Ref. [81]. Copyright (2015), ACS)

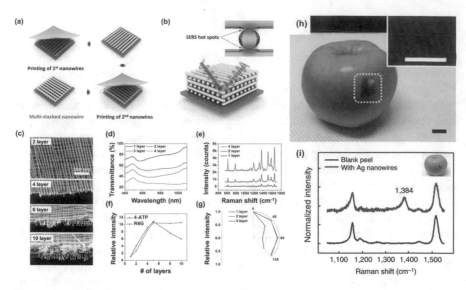

Fig. 11 (**a**) Fabrication process of 3D cross-point Au nanostructures. (**b**) Hot spot formation at cross-points. (**c**) SEM images of multi-stacking capabilities up to 10 layers. (**d**) Transmittance spectra of multi-stacked Au nanostructures. (**e**) SERS spectra of 3D cross-point Au nanostructures. (**f**) SERS intensity of R6G with increasing number of multi-stacked layers. (**g**) Angle dependency of multi-stacked Au nanostructures. (Reprinted with permission from Ref. [80]. Copyright (2016), Wiley.) (**h**) Nanotransfer printing of Ag nanostructures onto the surface of an apple. Scale bar 1 cm, 5 μm (inset). (**i**) SERS spectra of tetramethylthiuram disulfide (Thiram) using Ag nanostructures printed on the apple. (Reprinted with permission from Ref. [82]. Copyright (2014), Nature Publishing Group)

that form between rings significantly increases, leading to large Raman signal amplifications (Fig. 10o).

Nanotransfer printing is another powerful tool to fabricate plasmonic SERS structures. Previously reported plasmonic nanostructures are mostly based on two-dimensional patterns, yet increased hot spot density not only laterally but also vertically could further enhance Raman signal intensities. In this regard, 3D cross-point Au nanostructures have been produced for effective SERS applications [80] using nanotransfer printing [82]. As shown in Fig. 11a, b, Au nanowire layers are sequentially stacked upon each other with a 90° difference in orientation so that there are vertical hot spots between adjacent layers of Au nanowires. Multi-stacking capabilities up to 10 layers have been demonstrated, and Raman measurement results have shown maximum enhancement at 4–5 layers of multi-stacking due to limited penetration depths of excitation lasers (Fig. 11c–g). Nanotransfer printing is advantageous in forming plasmonic SERS structures on almost any desired surface. For example, Ag nanowires were printed onto the surface of an apple and Raman measurements were successfully taken (Fig. 11h, i).

4 Molecule Detection and Analysis

Rapid, facile, and noninvasive detection of trace-amount molecules is in demand for various fields of sensing applications [83]. Exploiting the advantages of Raman spectroscopy and low-cost, robust fabrication techniques of plasmonic SERS nanostructures, the requirements for highly sensitive trace-amount detection are expected to be satisfied. Major applications of SERS-based detection include single-molecule-level detection, multiplexing, biomedical sensing, and bio-imaging (Fig. 12).

In particular, SERS technologies are useful in biomedical applications for rapid and label-free identification of biomarkers and cancer cells, which enable effective diagnosis early into a disease [84]. For example, glucose levels in blood and human fluids are well known as an indicator of diabetes. A contact-lens-type SERS sensor for glucose in tears has been fabricated using Ag nanostructures and nanotransfer printing techniques [80]. As shown in Fig. 13a–d, Ag nanorod arrays were transfer-printed onto a hard and soft contact lens to form a "SERS contact lens." After the SERS contact lens was mounted on an artificial eye, detection of glucose levels as low as 10^{-4} M was successfully demonstrated (Fig. 13e). There still remain several challenges for practical applications such as the development of a safe excitation method and a data processing tool that enables quantitative analysis. Also, cancerous and noncancerous prostate cells within human fluids can be detected through Raman measurements. A detection platform using microfluidics and SERS has been fabricated for the analysis of mammalian cells flowing through human fluids [85]. Mixtures of cancerous and noncancerous cells were first incubated with SERS-sensitive tags for effective sensing. Hydrodynamic flow forces the cells to pass through the microfluidic channel in a single-file manner, allowing the Raman excitation laser to focus (Fig. 13f). Complex Raman spectra were processed using the classical least-squares (CLS) method for accurate analysis of cancerous and noncancerous cells (Fig. 13g, h). Although reliable in batch-to-batch stability, current Raman analysis technologies of living cells are limited to labelled sensing.

Furthermore, bio-imaging using SERS can effectively visualize the distribution of biomolecules within a living organism. In vitro bio-imaging of tumors has been achieved by fabricating plasmonic Raman tag particles [86]. Raman tags consist of mesoporous silica shells and sphere-shaped Au nanoparticles with an internal nanogap. As illustrated in Fig. 14a, in vivo Raman measurements were taken by extracting tumors from an orthotopic prostate cancer mouse after direct or

Fig. 12 Selected applications of SERS technologies

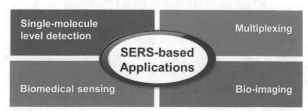

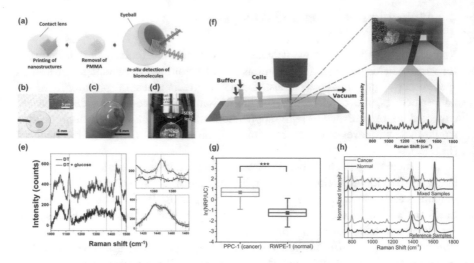

Fig. 13 (a) Fabrication of SERS contact lens. (b–d) Photograph of SERS contact lenses and Raman measurement of SERS contact lens mounted on artificial eye. (e) SERS spectra of glucose using SERS contact lens. (Reprinted with permission from Ref. [80]. Copyright (2016), Wiley.) (f) Schematic illustration of SERS measurement with microfluidic channel. (g, h) SERS spectra of cancerous and normal cells and CLS processing of SERS data. (Reprinted with permission from Ref. [85]. Copyright (2015), ACS)

intravenous injection of the Raman tags. Upon excitation, strong Raman signals were generated exclusively from tumors and remained stable for up to 30 min compared to unstable fluorescence images, which disappeared within 6 min. Also, SERS-based bio-imaging can enable vivid visualization of tumor margins or lesions in organs otherwise invisible to the naked eye. In vivo imaging of liver tumors was demonstrated using SERS nanoparticles containing Raman-sensitive reporter molecules [87]. SERS nanoparticles were Au cores coated with BPE encapsulated by a silica shell to prevent agglomeration and undesired contact with surroundings. Compared to fluorescence imaging, which uses indocyanine green (ICG), SERS nanoparticles showed higher contrast between normal and tumor cells in livers (Fig. 14b, c). The direct classical least-squares (DCLS) method aided the visualization of fluorescence and SERS imaging results.

5 Conclusion and Future Outlook

Detection of trace-amount molecules in natural form has been realized through various tools of analysis such as gas chromatography and mass spectrometry. However, due to the complexity in sample preparation and test methods, Raman spectroscopy has been suggested as an alternative, although weak signal intensity is the only drawback for wide applications. Recently, SERS, which utilizes plasmonic

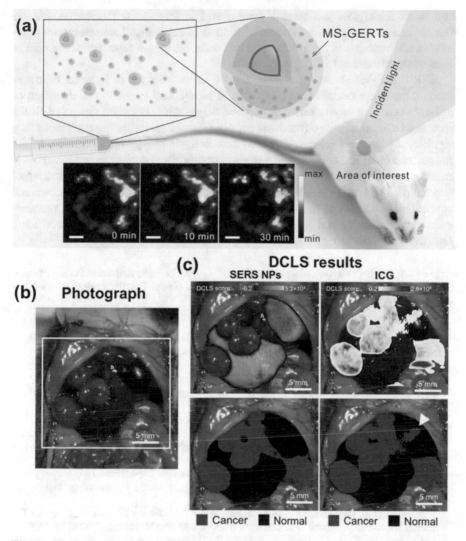

Fig. 14 (**a**) Bio-imaging using mesoporous silica-coated gap-enhanced Raman tags (MS-GERT) for visualization of tumors. (Reprinted with permission from Ref. [86]. Copyright (2017), ACS.) (**b, c**) Comparison of signal contrast for SERS-based bio-imaging and ICG-assisted fluorescence imaging of cancerous cells. (Reprinted with permission from Ref. [87]. Copyright (2016), ACS)

effects to amplify the Raman signal, has provided a solution. Since then, significant efforts have been devoted to developing plasmonic nanostructures for SERS, and thus a variety of fabrication techniques have been introduced.

Top–down methods such as e-beam lithography, nanoimprint lithography, and surface treatment are capable of precisely designing nanostructures, yet are costly. Bottom–up methods such as nanoparticle synthesis, self-assembly of nanoparticles, and block copolymer self-assembly are useful in large-scale fabrication, but are limited in reproducibility. Nanotransfer printing of noble metal nanostructures combines the advantages of both techniques and demonstrates excellent enhancement of Raman signals and notable reproducibility.

SERS analysis shows promising results in various fields in need of trace-amount molecule detection. In particular, in biomedical applications, detection of low-concentration biomarkers can improve diagnostics. Effective design of SERS structures combined with advances in portable Raman spectroscopy equipment may significantly improve in vivo detection and analysis of biological species. In the future, molecular and cellular SERS using nanofabricated structures and materials is expected to lead to breakthrough medical technology by early diagnosis, tracking of treatment effects, and simultaneous diagnosis and treatment. For more widespread applications of SERS in biomedical areas, however, achieving sufficient measurement reproducibility and selective detection of biomolecules without sacrificing sensitivity are remaining challenges.

Acknowledgments This work was supported by the Center for Integrated Smart Sensors, funded by the Ministry of Science, ICT and Future Planning as Global Frontier Project (CISS-2011-0031848).

References

1. Alvarez-Puebla R, Liz-Marzan L (2010) Environmental applications of plasmon assisted Raman scattering. Energy Environ Sci 3(8):1011–1017
2. Fainberg A (1992) Explosives detection for aviation security. Science 255(5051):1531
3. Wu X, Xu L, Liu L, Ma W, Yin H, Kuang H, Wang L, Xu C, Kotov NA (2013) Unexpected chirality of nanoparticle dimers and ultrasensitive chiroplasmonic bioanalysis. J Am Chem Soc 135(49):18629–18636
4. Rodríguez-Lorenzo L, De La Rica R, Álvarez-Puebla RA, Liz-Marzán LM, Stevens MM (2012) Plasmonic nanosensors with inverse sensitivity by means of enzyme-guided crystal growth. Nat Mater 11(7):604
5. Stern E, Vacic A, Rajan NK, Criscione JM, Park J, Ilic BR, Mooney DJ, Reed MA, Fahmy TM (2010) Label-free biomarker detection from whole blood. Nat Nanotechnol 5(2):138–142
6. Li G, Liu Y, Liu Y, Chen L, Wu S, Liu Y, Li X (2013) Photoaffinity labeling of small-molecule-binding proteins by DNA-templated chemistry. Angew Chem Int Ed 52(36):9544–9549
7. Choi Y, Park Y, Kang T, Lee LP (2009) Selective and sensitive detection of metal ions by plasmonic resonance energy transfer-based nanospectroscopy. Nat Nanotechnol 4(11):742–746
8. Cho ES, Kim J, Tejerina B, Hermans TM, Jiang H, Nakanishi H, Yu M, Patashinski AZ, Glotzer SC, Stellacci F (2012) Ultrasensitive detection of toxic cations through changes in the tunnelling current across films of striped nanoparticles. Nat Mater 11(11):978
9. Clarkson TW (2002) The three modern faces of mercury. Environ Health Perspect 110(Suppl 1):11
10. Onyido I, Norris AR, Buncel E (2004) Biomolecule—mercury interactions: modalities of DNA base—mercury binding mechanisms. Remediation strategies. Chem Rev 104(12):5911–5930
11. Shields PG (2006) Understanding population and individual risk assessment: the case of polychlorinated biphenyls. Cancer Epidemiol Biomarkers Prev 15:830–839

12. Mullins MD, Pochini CM, McCrindle S, Romkes M, Safe SH, Safe LM (1984) High-resolution PCB analysis: synthesis and chromatographic properties of all 209 PCB congeners. Environ Sci Technol 18(6):468–476
13. Matsumoto R, Tu NPC, Haruta S, Kawano M, Takeuchi I (2014) Polychlorinated biphenyl (PCB) concentrations and congener composition in masu salmon from Japan: a study of all 209 PCB congeners by high-resolution gas chromatography/high-resolution mass spectrometry (HRGC/HRMS). Mar Pollut Bull 85(2):549–557
14. Capriotti AL, Cavaliere C, Colapicchioni V, Piovesana S, Samperi R, Laganà A (2013) Analytical strategies based on chromatography–mass spectrometry for the determination of estrogen-mimicking compounds in food. J Chromatogr A 1313:62–77
15. Gao Y, Shi Z, Long Z, Wu P, Zheng C, Hou X (2012) Determination and speciation of mercury in environmental and biological samples by analytical atomic spectrometry. Microchem J 103:1–14
16. Giljohann DA, Mirkin CA (2009) Drivers of biodiagnostic development. Nature 462(7272):461
17. Cockerill F III, Wilson J, Vetter E, Goodman K, Torgerson C, Harmsen W, Schleck C, Ilstrup D, Washington JN, Wilson W (2004) Optimal testing parameters for blood cultures. Clin Infect Dis 38(12):1724–1730
18. Raman C, Krishnan K (1928) A new type of secondary radiation. Nature 121:501–502
19. Kneipp K, Kneipp H, Itzkan I, Dasari RR, Feld MS (1999) Ultrasensitive chemical analysis by Raman spectroscopy. Chem Rev 99(10):2957–2976
20. Casiraghi C, Pisana S, Novoselov K, Geim A, Ferrari A (2007) Raman fingerprint of charged impurities in graphene. Appl Phys Lett 91(23):233108
21. Hashimoto M, Araki T, Kawata S (2000) Molecular vibration imaging in the fingerprint region by use of coherent anti-Stokes Raman scattering microscopy with a collinear configuration. Opt Lett 25(24):1768–1770
22. Nie S, Emory SR (1997) Probing single molecules and single nanoparticles by surface-enhanced Raman scattering. Science 275(5303):1102–1106
23. Jeanmaire DL, Van Duyne RP (1977) Surface Raman spectroelectrochemistry: part I. Heterocyclic, aromatic, and aliphatic amines adsorbed on the anodized silver electrode. J Electroanal Chem Interfacial Electrochem 84(1):1–20
24. Fleischmann M, Hendra PJ, McQuillan AJ (1974) Raman spectra of pyridine adsorbed at a silver electrode. Chem Phys Lett 26(2):163–166
25. Moskovits M (1985) Surface-enhanced spectroscopy. Rev Mod Phys 57(3):783
26. Michaels AM, Jiang J, Brus L (2000) Ag nanocrystal junctions as the site for surface-enhanced Raman scattering of single rhodamine 6G molecules. J Phys Chem B 104(50):11965–11971
27. Xu H, Aizpurua J, Käll M, Apell P (2000) Electromagnetic contributions to single-molecule sensitivity in surface-enhanced Raman scattering. Phys Rev E Stat Phys Plasmas Fluids Relat Interdiscip Topics 62(3):4318–4324
28. Tian Z-Q, Ren B, Wu D-Y (2002) Surface-enhanced Raman scattering: from noble to transition metals and from rough surfaces to ordered nanostructures. J Phys Chem B 106:9463
29. Tao A, Kim F, Hess C, Goldberger J, He R, Sun Y, Xia Y, Yang P (2003) Langmuir–Blodgett silver nanowire monolayers for molecular sensing using surface-enhanced Raman spectroscopy. Nano Lett 3(9):1229–1233
30. Willets KA, Van Duyne RP (2007) Localized surface plasmon resonance spectroscopy and sensing. Annu Rev Phys Chem 58:267–297
31. Le F, Brandl DW, Urzhumov YA, Wang H, Kundu J, Halas NJ, Aizpurua J, Nordlander P (2008) Metallic nanoparticle arrays: a common substrate for both surface-enhanced Raman scattering and surface-enhanced infrared absorption. ACS Nano 2(4):707–718
32. García-Vidal FJ, Pendry J (1996) Collective theory for surface enhanced Raman scattering. Phys Rev Lett 77(6):1163
33. Moskovits M (1978) Surface roughness and the enhanced intensity of Raman scattering by molecules adsorbed on metals. J Chem Phys 69(9):4159–4161

34. Creighton JA, Blatchford CG, Albrecht MG (1979) Plasma resonance enhancement of Raman scattering by pyridine adsorbed on silver or gold sol particles of size comparable to the excitation wavelength. J Chem Soc Faraday Trans 75:790–798
35. Mulvihill MJ, Ling XY, Henzie J, Yang P (2009) Anisotropic etching of silver nanoparticles for plasmonic structures capable of single-particle SERS. J Am Chem Soc 132(1):268–274
36. Porter MD, Lipert RJ, Siperko LM, Wang G, Narayanan R (2008) SERS as a bioassay platform: fundamentals, design, and applications. Chem Soc Rev 37(5):1001–1011
37. Stiles PL, Dieringer JA, Shah NC, Van Duyne RP (2008) Surface-enhanced Raman spectroscopy. Annu Rev Anal Chem 1:601–626
38. Albrecht MG, Creighton JA (1977) Anomalously intense Raman spectra of pyridine at a silver electrode. J Am Chem Soc 99(15):5215–5217
39. Craig AP, Franca AS, Irudayaraj J (2013) Surface-enhanced Raman spectroscopy applied to food safety. Annu Rev Food Sci Technol 4:369–380
40. Yang W, Hulteen J, Schatz GC, Van Duyne RP (1996) A surface-enhanced hyper-Raman and surface-enhanced Raman scattering study of trans-1, 2-bis (4-pyridyl) ethylene adsorbed onto silver film over nanosphere electrodes. Vibrational assignments: experiment and theory. J Chem Phys 104(11):4313–4323
41. Kneipp K, Kneipp H, Itzkan I, Dasari RR, Feld MS (1999) Surface-enhanced Raman scattering: a new tool for biomedical spectroscopy. Curr Sci 77:915–924
42. Hao F, Sonnefraud Y, Van Dorpe P, Maier SA, Halas NJ, Nordlander P (2008) Symmetry breaking in plasmonic nanocavities: subradiant LSPR sensing and a tunable Fano resonance. Nano Lett 8(11):3983–3988. https://doi.org/10.1021/nl802509r
43. Kelly KL, Coronado E, Zhao LL, Schatz GC (2003) The optical properties of metal nanoparticles: the influence of size, shape, and dielectric environment. J Phys Chem B 107:668
44. Talley CE, Jackson JB, Oubre C, Grady NK, Hollars CW, Lane SM, Huser TR, Nordlander P, Halas NJ (2005) Surface-enhanced Raman scattering from individual Au nanoparticles and nanoparticle dimer substrates. Nano Lett 5(8):1569–1574
45. Lee J-H, Nam J-M, Jeon K-S, Lim D-K, Kim H, Kwon S, Lee H, Suh YD (2012) Tuning and maximizing the single-molecule surface-enhanced Raman scattering from DNA-tethered nanodumbbells. ACS Nano 6(11):9574–9584
46. Dieringer JA, Lettan RB, Scheidt KA, Van Duyne RP (2007) A frequency domain existence proof of single-molecule surface-enhanced Raman spectroscopy. J Am Chem Soc 129(51):16249–16256
47. Stokes RJ, Macaskill A, Lundahl PJ, Smith WE, Faulds K, Graham D (2007) Quantitative enhanced Raman scattering of labeled DNA from gold and silver nanoparticles. Small 3(9):1593–1601
48. Faulds K, Smith WE, Graham D (2004) Evaluation of surface-enhanced resonance Raman scattering for quantitative DNA analysis. Anal Chem 76(2):412–417
49. Campion A, Kambhampati P (1998) Surface-enhanced Raman scattering. Chem Soc Rev 27(4):241–250
50. Yang JK, Cord B, Duan H, Berggren KK, Klingfus J, Nam S-W, Kim K-B, Rooks MJ (2009) Understanding of hydrogen silsesquioxane electron resist for sub-5-nm-half-pitch lithography. J Vac Sci Technol 27(6):2622–2627
51. Feng L, Xu Y-L, Fegadolli WS, Lu M-H, Oliveira JE, Almeida VR, Chen Y-F, Scherer A (2013) Experimental demonstration of a unidirectional reflectionless parity-time metamaterial at optical frequencies. Nat Mater 12(2):108
52. Duan H, Hu H, Kumar K, Shen Z, Yang JK (2011) Direct and reliable patterning of plasmonic nanostructures with sub-10-nm gaps. ACS Nano 5(9):7593–7600
53. Gopalakrishnan A, Chirumamilla M, De Angelis F, Toma A, Zaccaria RP, Krahne R (2014) Bimetallic 3D nanostar dimers in ring cavities: recyclable and robust surface-enhanced Raman scattering substrates for signal detection from few molecules. ACS Nano 8(8):7986–7994
54. Boyack R, Le Ru EC (2009) Investigation of particle shape and size effects in SERS using T-matrix calculations. Phys Chem Chem Phys 11(34):7398–7405

55. Kim A, Ou FS, Ohlberg DA, Hu M, Williams RS, Li Z (2011) Study of molecular trapping inside gold nanofinger arrays on surface-enhanced Raman substrates. J Am Chem Soc 133(21):8234–8239
56. Lee SY, Kim S-H, Kim MP, Jeon HC, Kang H, Kim HJ, Kim BJ, Yang S-M (2013) Freestanding and arrayed nanoporous microcylinders for highly active 3D SERS substrate. Chem Mater 25(12):2421–2426
57. Huang J-A, Zhao Y-Q, Zhang X-J, He L-F, Wong T-L, Chui Y-S, Zhang W-J, Lee S-T (2013) Ordered Ag/Si nanowires array: wide-range surface-enhanced Raman spectroscopy for reproducible biomolecule detection. Nano Lett 13(11):5039–5045
58. Wu K, Rindzevicius T, Schmidt MS, Mogensen KB, Hakonen A, Boisen A (2015) Wafer-scale leaning silver nanopillars for molecular detection at ultra-low concentrations. J Phys Chem C 119(4):2053–2062
59. Jackson JB, Halas NJ (2004) Surface-enhanced Raman scattering on tunable plasmonic nanoparticle substrates. Proc Natl Acad Sci U S A 101(52):17930–17935
60. Wei W, Li S, Millstone JE, Banholzer MJ, Chen X, Xu X, Schatz GC, Mirkin CA (2009) Surprisingly long-range surface-enhanced Raman scattering (SERS) on Au–Ni multisegmented nanowires. Angew Chem Int Ed 48(23):4210–4212
61. Sanz-Ortiz MN, Sentosun K, Bals S, Liz-Marzán LM (2015) Templated growth of surface enhanced Raman scattering-active branched gold nanoparticles within radial mesoporous silica shells. ACS Nano 9(10):10489–10497
62. Kang JW, So PT, Dasari RR, Lim D-K (2015) High resolution live cell Raman imaging using subcellular organelle-targeting SERS-sensitive gold nanoparticles with highly narrow intra-nanogap. Nano Lett 15(3):1766–1772
63. Kumar PS, Pastoriza-Santos I, Rodriguez-Gonzalez B, de Abajo FJG, Liz-Marzan LM (2007) High-yield synthesis and optical response of gold nanostars. Nanotechnology 19(1):015606
64. Peng B, Li G, Li D, Dodson S, Zhang Q, Zhang J, Lee YH, Demir HV, Yi Ling X, Xiong Q (2013) Vertically aligned gold nanorod monolayer on arbitrary substrates: self-assembly and femtomolar detection of food contaminants. ACS Nano 7(7):5993–6000
65. Taylor RW, Coulston RJ, Biedermann F, Mahajan S, Baumberg JJ, Scherman OA (2013) In situ SERS monitoring of photochemistry within a nanojunction reactor. Nano Lett 13(12):5985–5990
66. Taylor RW, Esteban RN, Mahajan S, Coulston R, Scherman OA, Aizpurua J, Baumberg JJ (2012) Simple composite dipole model for the optical modes of strongly-coupled plasmonic nanoparticle aggregates. J Phys Chem C 116(47):25044–25051
67. Ni H, Wang M, Shen T, Zhou J (2015) Self-assembled large-area annular cavity arrays with tunable cylindrical surface plasmons for sensing. ACS Nano 9(2):1913–1925
68. Lee J, Zhang Q, Park S, Choe A, Fan Z, Ko H (2015) Particle–film plasmons on periodic silver film over nanosphere (AgFON): a hybrid plasmonic nanoarchitecture for surface-enhanced Raman spectroscopy. ACS Appl Mater Interfaces 8(1):634–642
69. Poujet Y, Salvi J, Baida FI (2007) 90% extraordinary optical transmission in the visible range through annular aperture metallic arrays. Opt Lett 32(20):2942–2944
70. Park M, Harrison C, Chaikin PM, Register RA, Adamson DH (1997) Block copolymer lithography: periodic arrays of~ 10 11 holes in 1 square centimeter. Science 276(5317):1401–1404
71. Vega DA, Harrison CK, Angelescu DE, Trawick ML, Huse DA, Chaikin PM, Register RA (2005) Ordering mechanisms in two-dimensional sphere-forming block copolymers. Phys Rev E Stat Nonlin Soft Matter Phys 71(6):061803
72. Cheng JY, Ross C, Chan VH, Thomas EL, Lammertink RG, Vancso GJ (2001) Formation of a cobalt magnetic dot array via block copolymer lithography. Adv Mater 13(15):1174–1178
73. Cha SK, Mun JH, Chang T, Kim SY, Kim JY, Jin HM, Lee JY, Shin J, Kim KH, Kim SO (2015) Au–Ag core–shell nanoparticle array by block copolymer lithography for synergistic broadband plasmonic properties. ACS Nano 9(5):5536–5543

74. Yap FL, Thoniyot P, Krishnan S, Krishnamoorthy S (2012) Nanoparticle cluster arrays for high-performance SERS through directed self-assembly on flat substrates and on optical fibers. ACS Nano 6(3):2056–2070
75. Lee W, Lee SY, Briber RM, Rabin O (2011) Self-assembled SERS substrates with tunable surface plasmon resonances. Adv Funct Mater 21(18):3424–3429
76. Jung YS, Ross CA (2007) Orientation-controlled self-assembled nanolithography using a polystyrene—polydimethylsiloxane block copolymer. Nano Lett 7(7):2046–2050
77. Sundrani D, Darling S, Sibener S (2004) Hierarchical assembly and compliance of aligned nanoscale polymer cylinders in confinement. Langmuir 20(12):5091–5099
78. Sundrani D, Sibener S (2002) Spontaneous spatial alignment of polymer cylindrical nanodomains on silicon nitride gratings. Macromolecules 35(22):8531–8539
79. Black C (2005) Self-aligned self assembly of multi-nanowire silicon field effect transistors. Appl Phys Lett 87(16):163116
80. Jeong JW, Arnob MMP, Baek KM, Lee SY, Shih WC, Jung YS (2016) 3D cross-point plasmonic nanoarchitectures containing dense and regular hot spots for surface-enhanced Raman spectroscopy analysis. Adv Mater 28(39):8695–8704
81. Baek KM, Kim JM, Jeong JW, Lee SY, Jung YS (2015) Sequentially self-assembled rings-in-mesh nanoplasmonic arrays for surface-enhanced Raman spectroscopy. Chem Mater 27(14):5007–5013
82. Jeong JW, Yang SR, Hur YH, Kim SW, Baek KM, Yim S, Jang H-I, Park JH, Lee SY, Park C-O (2014) High-resolution nanotransfer printing applicable to diverse surfaces via interface-targeted adhesion switching. Nat Commun 5:5387
83. Zielinski O, Busch JA, Cembella AD, Daly KL, Engelbrektsson J, Hannides AK, Schmidt H (2009) Detecting marine hazardous substances and organisms: sensors for pollutants, toxins, and pathogens. Ocean Sci 5:329
84. Ullal AV, Peterson V, Agasti SS, Tuang S, Juric D, Castro CM, Weissleder R (2014) Cancer cell profiling by barcoding allows multiplexed protein analysis in fine-needle aspirates. Sci Transl Med 6(219):219ra9
85. Pallaoro A, Hoonejani MR, Braun GB, Meinhart CD, Moskovits M (2015) Rapid identification by surface-enhanced Raman spectroscopy of cancer cells at low concentrations flowing in a microfluidic channel. ACS Nano 9(4):4328–4336
86. Zhang Y, Qiu Y, Lin L, Gu H, Xiao Z, Ye J (2017) Ultraphotostable mesoporous silica-coated gap-enhanced Raman tags (GERTs) for high-speed bioimaging. ACS Appl Mater Interfaces 9(4):3995–4005
87. Andreou C, Neuschmelting V, Tscharganeh D-F, Huang C-H, Oseledchyk A, Iacono P, Karabeber H, Colen RR, Mannelli L, Lowe SW (2016) Imaging of liver tumors using surface-enhanced Raman scattering nanoparticles. ACS Nano 10(5):5015–5026

Basics and Advances in Monocular vSLAM

Hideaki Uchiyama, Takafumi Taketomi, Sei Ikeda, and Shohei Mori

1 Introduction

Augmented reality (AR) is a research field dealing with the enhancement of human perception with information and communication technologies. Since 1990s, visual enhancement has been proposed such that users see an object in the 3D space through a display with a camera such as a mobile phone, and computer-generated (CG) objects are seamlessly superimposed onto images captured by the camera [2]. The main motivation of the enhancement is to support human activities by visualizing instructive annotations. This visualization technology has widely been applied to user interfaces and image synthesis for broadcasting [13]. In contrast to AR based visualization adding CG objects onto images, diminished reality (DR) has also been proposed as a new concept that unneeded or undesirable objects in the images are naturally hidden or removed by visualizing their background, as an operation of subtraction [29]. Nowadays, other perceptions such as taste sensation have also been enhanced, and such technologies can be categorized into a new

H. Uchiyama (✉)
Kyushu University, Fukuoka, Japan
e-mail: uchiyama@limu.ait.kyushu-u.ac.jp

T. Taketomi
Nara Institute of Science and Technology, Nara, Japan
e-mail: takafumi-t@is.naist.jp

S. Ikeda
Ritsumeikan University, Kyoto, Japan
e-mail: ikeda.sei.jp@ieee.org

S. Mori
Keio University, Tokyo, Japan
e-mail: s.mori.jp@ieee.org

© Springer Nature Switzerland AG 2020
Y. Liu et al. (eds.), *Smart Sensors and Systems*,
https://doi.org/10.1007/978-3-030-42234-9_5

research field referred to as augmented human [32, 36]. In this chapter, we focus on the traditional AR that enhances human vision using computer vision technologies.

To develop AR applications, one important technology is to estimate camera poses of sequential images with respect to an object in the 3D space [27]. A camera pose is generally represented by a 3D position and a 3D orientation with respect to a certain coordinate system. In other words, it is equivalent to 6 Degree-of-Freedom (DoF) parameterization. Camera pose estimation is crucial for continuously superimposing CG objects onto the images as if they really exist in the 3D space. Specifically, it is important to keep geometric and photometric consistencies between the CG objects and the 3D space. This technical issue is generally referred to as visual tracking because camera poses are computed and tracked only from visual information. If IMU is available for the tracking, the issue can be categorized into visual-inertial tracking [35].

Existing visual tracking algorithms can be divided into two main approaches: model-based and model-less ones. In the model-based approaches, the 3D model of an object is given or generated beforehand as a reference, and camera poses are computed with respect to the object using the model [20]. The model normally comprises visual features such as a set of 3D points or lines with textures. Since such model is not always available, model-less approaches have also been investigated, and are generally referred to as visual simultaneous localization and mapping (vSLAM) [40]. In the vSLAM, camera pose estimation (localization) and online 3D model generation (mapping) are iteratively performed so that camera poses can sequentially be computed in unprepared environments. The vSLAM has been considered an active and important research topic in computer vision, robotics, and augmented reality for decades [3, 11]. Since the performance of vSLAM has been drastically improved, vSLAM can run in real time even on mobile devices.

In this chapter, we review visual tracking technologies for augmented reality, computer vision, and robotics applications. Specifically, we focus on vSLAM algorithms using a monocular RGB camera as a model-less approach that has actively been investigated in recent years. The chapter comprises two main sections: basics and advances of the vSLAM. In the basics, basic computer vision technologies used in vSLAM are first explained because they are necessary for understanding the process of vSLAM algorithms. Then, a basic framework and existing vSLAM algorithms are summarized. This paper aims to explain vSLAM algorithms for non-experts who started to learn the algorithms. Note that vSLAM using a RGB-D camera can be found [4, 16].

2   Basics

The vSLAM can be achieved with the combination of several computer vision technologies. To understand vSLAM algorithms, the knowledge of each technology is required. First, camera geometry, keypoint matching, triangulation, and bundle adjustment are introduced as basics of the vSLAM.

2.1 Camera Geometry

As illustrated in Fig. 1, a point in the world coordinate system $X_w = [X_w, Y_w, Z_w]^T$ is projected onto the image coordinate system $u = [u, v]^T$ by a perspective projection model:

$$s \begin{bmatrix} u \\ v \\ 1 \end{bmatrix} = \begin{bmatrix} f_u & 0 & c_u \\ 0 & f_v & c_v \\ 0 & 0 & 1 \end{bmatrix} \begin{bmatrix} r_{11} & r_{12} & r_{13} & t_1 \\ r_{21} & r_{22} & r_{23} & t_2 \\ r_{31} & r_{32} & r_{33} & t_3 \end{bmatrix} \begin{bmatrix} X_w \\ Y_w \\ Z_w \\ 1 \end{bmatrix}$$

$$s\tilde{u} = K \begin{bmatrix} r_1 & r_2 & r_3 & t \end{bmatrix} \tilde{X}_w$$

$$= K \begin{bmatrix} R & t \end{bmatrix} \tilde{X}_w$$

$$= P\tilde{X}_w,$$

(1)

where f_u, f_v, c_u, and c_v are focal lengths and the principal point as camera intrinsic parameters, K is a matrix containing the intrinsic parameters, r_{ij} and t_i are the elements of the 3×3 rotation matrix R and the 3×1 translation vector t as camera extrinsic parameters, P is the 3×4 projection matrix containing both camera intrinsic and extrinsic parameters, $\tilde{\ }$ is a homogeneous coordinate that is useful for matrix operations in projective geometry [14], and s is a scale factor. t represents the origin of the world coordinate system in the camera coordinate system, and r_1, r_2, and r_3 are the basis vectors of the world coordinate system in the camera coordinate system. In Eq. (1), a lens distortion effect is not considered for a brief description. In general, the distortion process is applied after u is computed [14].

Since camera intrinsic parameters including lens distortion parameters can be fixed when a lens and an image sensor are determined, they are computed beforehand by a camera calibration method [43], and are considered as known parameters. Therefore, camera pose estimation in the vSLAM is equivalent to computing R and t from sets of X_w and u. This issue is referred to as the Perspective-n-Point (PnP) problem, and several solutions have been proposed [21, 23]. In the next section, the

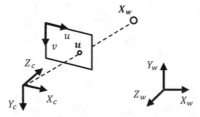

Fig. 1 Coordinate systems for vSLAM. The world coordinate system X_w is first defined in 3D space. The camera coordinate system X_c is then defined such that $X_c Y_c$ plane is parallel to the image coordinate system u and Z_c axis is the optical axis

process of computing X_w and u with keypoint matching is explained. It should be noted that the relationship between the world coordinate system X_w and the camera coordinate system X_c is represented by the following equation:

$$X_c = RX_w + t$$

$$\tilde{X}_c = \begin{bmatrix} R & t \\ 0^T & 1 \end{bmatrix} \tilde{X}_w. \tag{2}$$

2.2 Keypoint Matching

To compute R and t in Eq. (1), it is necessary to compute the correspondences between X_w and u. Since X_w is given when a 3D model is available, the issue is how to detect a pixel u in an image projected from a 3D point X_w. This can normally be done by using keypoint matching [15, 28].

Figure 2 illustrates an example of camera pose estimation with respect to a box. The process of keypoint matching is divided into extraction, description, and matching. The texture and shape of a 3D model are first prepared as a reference where the world coordinate system is defined. In the extraction, pixels, where local intensity distribution around the pixels are different from the others or discriminative, are selected in both the texture of the 3D model and the input image for camera pose estimation. Such pixels are referred to as feature points or keypoints [22, 42]. In the description, the canonical orientation of a keypoint is

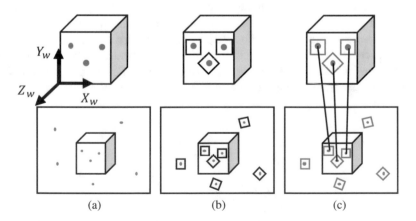

$$(a) \qquad\qquad (b) \qquad\qquad (c)$$

Fig. 2 Keypoint matching. The first row represents the process for a 3D model and the second one does that for the input image. Keypoint matching is composed of (**a**) extraction of keypoints, feature (**b**) description for each keypoint, and (**c**) matching of keypoints between the model and the input image

first computed from the intensity distribution [12]. This process allows keypoint matching to be invariant to rotational changes of the images. The feature vector of a keypoint is then computed from the intensity distribution with a feature descriptor such as the histogram of gradients. In the literature, various methods have been proposed for these two processes [26, 38]. In the matching, the feature vector of a keypoint in the image is matched with those in the reference. This is generally solved by using a nearest neighbor searching method [1].

From keypoint matching, the correspondences between X_w and u can be computed and are used for computing R and t in Eq. (1) with the solutions for the PnP problem [21, 23]. Since the correspondences may include some wrong ones, a robust estimator is incorporated in the computation of R and t [5, 9].

2.3 3D Reconstruction

In the model-based visual tracking, the 3D model of an object must be generated offline, and camera poses are computed with respect to the object online. On the other hand, vSLAM algorithms compute the 3D model of an object or 3D space online in addition to camera pose estimation. This process is referred to as 3D reconstruction or mapping.

The simplest mapping method is the triangulation using two views. Figure 3 illustrates an example of the triangulation for a point X_w observed in the two images whose projection matrices are P and P', and the projected locations in the image coordinate system are u and u', respectively. From Eq. (1), the projection to each image can be represented as follows:

$$s\tilde{u} = P\tilde{X}_w$$
$$s'\tilde{u}' = P'\tilde{X}_w. \tag{3}$$

In the two equations, u, u', P, and P' must be known such that sets of u and u' can be computed by using keypoint matching between the two images, and P and P' are

Fig. 3 Triangulation. X_w is computed by solving Eq. (3) with u, u', P, and P'

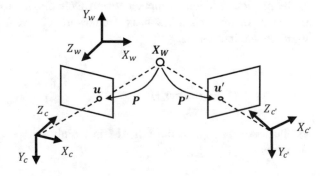

computed from a reference or objects whose 3D models are known in the 3D space. In other words, only X_w is unknown parameters. Since the degree of freedom of X_w is 3 and the number of equations is 4, X_w can be computed with a least squares solution [14].

2.4 Bundle Adjustment

In the vSLAM, camera poses are estimated from objects whose 3D models are known, and the models of new parts in the space are reconstructed to allow them to be known. In this process, the error of camera pose estimation and 3D reconstruction can be accumulated. To globally refine all of the estimated results, bundle adjustment is normally applied [25, 41]. First, Eq. (1) is expressed without using a homogeneous coordinate as follows:

$$u = \text{proj}(K, R, t, X), \tag{4}$$

where proj is a function to project a 3D point X onto the image coordinate system using K, R, and t. Then, the following energy function for minimizing reprojection error is defined:

$$E = \sum_i \sum_{j \in F_i} \| u_j - \text{proj}\left(K, R_i, t_i, X_j\right) \|^2, \tag{5}$$

where i is a number for images, j is a number for 3D points, and F_i is a set of points visible in an image i. By updating all of the R_i, t_i, and X_j, the energy is minimized. Since this optimization process is highly non-linear, a local minimum can be computed.

3 Advances

In the previous section, computer vision technologies used in the vSLAM were explained. In this section, a basic framework of vSLAM and existing vSLAM algorithm are introduced.

3.1 vSLAM Framework

As illustrated in Fig. 4, the vSLAM is mainly composed of the following five processes.

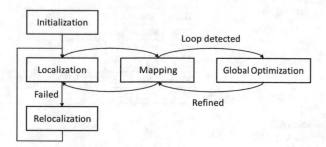

Fig. 4 vSLAM framework. vSLAM is mainly composed of five processes: initialization, localization, mapping, relocalization, and global optimization

- Initialization
- Localization (equivalent to camera pose estimation)
- Mapping (equivalent to 3D reconstruction)
- Relocalization
- Global optimization (bundle adjustment, etc.) by loop closure

The initialization is the process for both defining the world coordinate system and generating an initial 3D model in the coordinate system. This is necessary at the beginning of the vSLAM because the world coordinate system must be defined to compute camera poses. For instance, a fiducial marker can be prepared in the 3D space for the initialization [6]. After the initialization, the localization and mapping are performed. The relocalization is the process for computing a camera pose when the localization is failed due to fast camera motion or some disturbances. The global optimization is the process for refining all of the results of both camera pose estimation and 3D reconstruction. The bundle adjustment is one method for the optimization. The optimization effectively works if a re-visiting moment when the same object is captured again is detected to generate a closed loop. To understand vSLAM algorithms, it is important to know how each process is achieved.

3.2 History

The history of vSLAM algorithms is summarized in Table 1, and their relationship is illustrated in Fig. 5. The algorithms are categorized from two aspects: localization and mapping methods, and density of 3D reconstruction. As a framework of the localization and mapping, feature based methods have been proposed since 2003 [6, 18, 31]. In these methods, keypoints are extracted in the images, and used for the localization and mapping. Therefore, objects containing keypoints must exist in the environments. Since the drawback of these methods is that the methods do not work under texture-less environments, direct methods using more pixels in the image, namely feature-less methods because feature points are explicitly not extracted, have

Table 1 History of vSLAM
algorithms

Year	Algorithm
2003	MonoSLAM [6]
2007	PTAM [18]
2011	DTAM [33]
2014	LSD-SLAM [7], SVO [10], ORB-SLAM [31]
2017	DSO [8]

Fig. 5 Classification of
vSLAM algorithms. vSLAM
algorithms are classified
according to localization and
mapping methods
(direct/feature based), and the
density of 3D reconstruction
(dense/sparse)

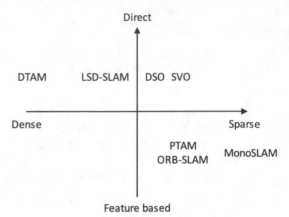

been proposed to allow vSLAM to be more robust, as an alternative framework [7, 33]. The density of 3D reconstruction is also an important aspect. Basically, the density of 3D reconstruction in feature based methods can be determined from the number of keypoints extracted in image sequences. Normally, this density is referred to as sparse. On the other hand, the density in direct methods varies according to the methods. For example, the map of the DTAM is highly dense because all of the pixels in the image are used for 3D reconstruction [33]. Then, the density has become more sparse [8, 10].

It should be noted that one of the drawbacks of feature based approaches is the sparsity of points in the 3D reconstruction. This indicates that a robust feature selection regarding appearance changes due to viewpoint changes is the minimum requirement to compute camera poses robustly and accurately. However, in terms of 3D reconstruction for scene understanding and photometric registration in AR environments, fully reconstructed 3D space is preferable. Therefore, RGB-D SLAM that achieves dense reconstruction is often used for solving the issues [24, 37].

3.3   Feature Based Approach

The world's first monocular 6DoF vSLAM was proposed in 2003 by Andrew Davison [6], namely MonoSLAM. MonoSLAM achieved the localization and mapping using extended Kalman filtering (EKF). A camera pose and 3D positions of keypoints are represented as a state vector in EKF. As camera moves, new keypoints are added into the state vector. Therefore, target environments were relatively small because the state vector becomes larger and cannot be updated in real time as the environment becomes larger.

To solve the computational cost issue for larger environments in MonoSLAM, a novel framework spiting the localization and mapping tasks into different threads on CPU was proposed in PTAM [18]. The advantage of this framework is that the computational cost of the mapping does not affect the localization because the localization and mapping are performed in different threads. As a result, bundle adjustment was incorporated into the mapping to refine the 3D reconstruction. This indicates that the localization thread estimates camera poses in real time, and the mapping thread performs accurate 3D reconstruction using bundle adjustment. After PTAM was proposed, many vSLAM algorithms based on multi-threading approaches have been proposed for real-time applications.

ORB-SLAM [31] can be considered an extended PTAM for large scale environments. If the error accumulation becomes larger as a camera moves, it may be difficult to compute the global minimum in bundle adjustment due to the numerous number of parameters in camera poses and keypoints in the 3D space. Pose graph optimization is a solution to avoid this problem [19]. After a loop is detected at a re-visiting moment [30], only camera poses are first optimized using the loop constraint. Then, bundle adjustment is performed to refine the results of both camera poses and 3D reconstruction.

The feature based approaches are finally summarized as follows:

- MonoSLAM: EKF based vSLAM
- PTAM: Keyframe based vSLAM
- ORB-SLAM: PTAM + Pose graph optimization

3.4 Direct Approach

In the feature based approaches, only geometric consistency such as keypoint positions in an image and 3D space are considered for the localization and mapping. In contrast, direct approaches are a new framework incorporating photometric consistency such that a camera pose is computed by comparing the image intensity with the intensity of keypoints in the 3D model.

DTAM is the first direct method for the vSLAM. The localization is done by comparing the input image with synthetic view images generated from the 3D model. This process can simply be considered registration between two images [33]. Since the synthesis of view images is time-consuming, it is implemented on GPU. The mapping is based on using multi-baseline stereo [34] for all of the pixels, and then it is refined using a surface continuity [39].

In DTAM, all of the pixels in an image are reconstructed. However, they are redundant for the localization task. Therefore, the mapping of the areas containing high intensity gradient was proposed in LSD-SLAM [7]. The idea of this approach is to ignore texture-less areas because it is difficult to estimate accurate depth information from images. In addition, loop closure detection and pose graph optimization were incorporated.

SVO [10] and DSO [8] are the direct method for sparse keypoints. In SVO, the localization is based on a feature based approach while the mapping is done with the direct method. DSO is a fully direct method such that the input image is divided into several blocks, and then high intensity points are selected for the localization and mapping. Since the direct method can be affected by photometric changes such as illumination, photometric camera calibration is performed online.

The direct approaches are finally summarized as follows:

- DTAM: Dense mapping with all the pixels
- LSD-SLAM: Semi-dense mapping for pixels containing high intensity gradients
- SVO and DSO: Sparse mapping with keypoints

4 Conclusion

This paper presented basics and recent advances of vSLAM as a framework of visual tracking in unprepared environments. Since the vSLAM can be achieved with the combination of several computer vision technologies, camera geometry, keypoint matching, triangulation, and bundle adjustment were introduced as basics. In the vSLAM, a common basic framework is composed of initialization, localization, mapping, relocalization, and global optimization with loop closure. For the localization and mapping, there are two main approaches: feature based and direct ones, and their history was introduced. The framework described in this paper assumes that an image is captured with global shutter indicating all of the pixels are captured at the same moment. However, rolling shutter is normally used for CMOS cameras and does not preserve the camera geometry in Eq. (1). This indicates that all of the pixels are captured at different moments and this should be considered in the localization and mapping [17].

References

1. Arya S, Mount DM, Netanyahu NS, Silverman R, Wu AY (1998) An optimal algorithm for approximate nearest neighbor searching fixed dimensions. J ACM 45(6):891–923
2. Azuma RT (1997) A survey of augmented reality. Presence Teleop Virt 6(4):355–385
3. Bresson G, Alsayed Z, Yu L, Glaser S (2017) Simultaneous localization and mapping: a survey of current trends in autonomous driving. In: IEEE transactions on intelligent vehicles
4. Chen K, Lai Y-K, Hu S-M (2015) 3d indoor scene modeling from RGB-D data: a survey. Computat Vis Media 1(4):267–278
5. Chum O, Matas J (2005) Matching with PROSAC-progressive sample consensus. In: IEEE computer society conference on computer vision and pattern recognition, 2005. CVPR 2005, vol 1. IEEE, Piscataway, pp 220–226

6. Davison AJ (2003) Real-time simultaneous localisation and mapping with a single camera. In: null. IEEE, Piscataway, p 1403
7. Engel J, Schöps T, Cremers D (2014) LSD-SLAM: large-scale direct monocular SLAM. In: European conference on computer vision. Springer, Berlin, pp 834–849
8. Engel J, Koltun V, Cremers D (2017) Direct sparse odometry. In: IEEE transactions on pattern analysis and machine intelligence
9. Fischler MA, Bolles RC (1981) Random sample consensus: a paradigm for model fitting with applications to image analysis and automated cartography. Commun ACM 24(6):381–395
10. Forster C, Pizzoli M, Scaramuzza D (2014) SVO: fast semi-direct monocular visual odometry. In: 2014 IEEE international conference on robotics and automation (ICRA). IEEE, Piscataway, pp 15–22
11. Fuentes-Pacheco J, Ruiz-Ascencio J, Rendón-Mancha JM (2015) Visual simultaneous localization and mapping: a survey. Artif Intell Rev 43(1):55–81
12. Gauglitz S, Turk M, Höllerer T (2011) Improving keypoint orientation assignment. In: Proceedings of BMVC, pp 1–11
13. Han J, Farin D et al (2007) A real-time augmented-reality system for sports broadcast video enhancement. In: Proceedings of the 15th ACM international conference on multimedia. ACM, New York, pp 337–340
14. Hartley R, Zisserman A (2003) Multiple view geometry in computer vision. Cambridge University Press, Cambridge
15. Heinly J, Dunn E, Frahm J-M (2012) Comparative evaluation of binary features. In: Computer vision–ECCV 2012. Springer, Berlin, pp 759–773
16. Hitomi EE, Silva JV, Ruppert GC (2015) 3d scanning using RGBD imaging devices: a survey. In: Developments in medical image processing and computational vision. Springer, Berlin, pp 379–395
17. Kim J-H, Cadena C, Reid I (2016) Direct semi-dense SLAM for rolling shutter cameras. In: 2016 IEEE international conference on robotics and automation (ICRA). IEEE, Piscataway, pp 1308–1315
18. Klein G, Murray D (2007) Parallel tracking and mapping for small AR workspaces. In: 6th IEEE and ACM international symposium on mixed and augmented reality, 2007. ISMAR 2007. IEEE, Piscataway, pp 225–234
19. Kümmerle R, Grisetti G, Strasdat H, Konolige K, Burgard W (2011) g2o: a general framework for graph optimization. In: 2011 IEEE international conference on robotics and automation (ICRA). IEEE, Piscataway, pp 3607–3613
20. Lepetit V, Fua P et al (2005) Monocular model-based 3d tracking of rigid objects: a survey. Found Trends Comput Graph Vis 1(1):1–89
21. Lepetit V, Moreno-Noguer F, Fua P (2009) EPnP: An accurate O(n) solution to the PnP problem. Int J Comput Vis 81(2):155–166
22. Li J, Allinson NM (2008) A comprehensive review of current local features for computer vision. Neurocomputing 71(10):1771–1787
23. Li S, Xu C, Xie M (2012) A robust O (n) solution to the perspective-n-point problem. IEEE Trans Pattern Anal Mach Intell 34(7):1444–1450
24. Li C, Xiao H, Tateno K, Tombari F, Navab N, Hanger GD (2016) Incremental scene understanding on dense SLAM. In: 2016 IEEE/RSJ international conference on intelligent robots and systems (IROS). IEEE, Piscataway, pp 574–581
25. Lourakis MI, Argyros AA (2009) SBA: a software package for generic sparse bundle adjustment. ACM Trans Math Softw 36(1):2
26. Lowe DG (2004) Distinctive image features from scale-invariant keypoints. Int J Comput Vis 60(2):91–110
27. Marchand E, Uchiyama H, Spindler F (2016) Pose estimation for augmented reality: a hands-on survey. IEEE Trans Vis Comput Graph 22(12):2633–2651
28. Miksik O, Mikolajczyk K (2012) Evaluation of local detectors and descriptors for fast feature matching. In: 2012 21st international conference on pattern recognition (ICPR). IEEE, Piscataway, pp 2681–2684

29. Mori S, Ikeda S, Saito H (2017) A survey of diminished reality: techniques for visually concealing, eliminating, and seeing through real objects. IPSJ Trans Comput Vis Appl 9(1):17
30. Mur-Artal R, Tardós JD (2014) Fast relocalisation and loop closing in keyframe-based slam. In: 2014 IEEE international conference on robotics and automation (ICRA). IEEE, Piscataway, pp 846–853
31. Mur-Artal R, Tardós JD (2014) ORB-SLAM: tracking and mapping recognizable features. In: MVIGRO workshop at robotics science and systems (RSS), Berkeley
32. Nakamura H, Miyashita H (2011) Augmented gustation using electricity. In: Proceedings of the 2nd augmented human international conference. ACM, New York, p 34
33. Newcombe RA, Lovegrove SJ, Davison AJ (2011) DTAM: Dense tracking and mapping in real-time. In: 2011 IEEE international conference on computer vision (ICCV). IEEE, Piscataway, pp 2320–2327
34. Okutomi M, Kanade T (1993) A multiple-baseline stereo. IEEE Trans Pattern Anal Mach Intell 15(4):353–363
35. Porzi L, Ricci E, Ciarfuglia TA, Zanin M (2012) Visual-inertial tracking on android for augmented reality applications. In: 2012 IEEE workshop on environmental energy and structural monitoring systems (EESMS). IEEE, Piscataway, pp 35–41
36. Ranasinghe N, Suthokumar G, Lee K-Y, Do EY-L (2015) Digital flavor: towards digitally simulating virtual flavors. In: Proceedings of the 2015 ACM on international conference on multimodal interaction. ACM, New York, pp 139–146
37. Richter-Trummer T, Kalkofen D, Park J, Schmalstieg D (2016) Instant mixed reality lighting from casual scanning. In: 2016 IEEE international symposium on mixed and augmented reality (ISMAR). IEEE, Piscataway, pp 27–36
38. Rublee E, Rabaud V, Konolige K, Bradski G (2011) ORB: an efficient alternative to sift or surf. In: 2011 IEEE international conference on computer vision (ICCV). IEEE, Piscataway, pp 2564–2571
39. Rudin LI, Osher S, Fatemi E (1992) Nonlinear total variation based noise removal algorithms. Physica D 60(1–4):259–268
40. Taketomi T, Uchiyama H, Ikeda S (2017) Visual slam algorithms: a survey from 2010 to 2016. IPSJ Trans Comput Vis Appl 9(1):16
41. Triggs B, McLauchlan PF, Hartley RI, Fitzgibbon AW (1999) Bundle adjustment a modern synthesis. In: International workshop on vision algorithms. Springer, Berlin, pp 298–372
42. Tuytelaars T, Mikolajczyk K et al (2008) Local invariant feature detectors: a survey. Found Trends Comput Graph Vis 3(3):177–280
43. Zhang Z (2000) A flexible new technique for camera calibration. IEEE Trans Pattern Anal Mach Intell 22(11):1330–1334

Wireless Backhaul Technology in Wireless Sensor Networks: Basic Framework and Applications

Hiroaki Togashi, Ryuta Abe, Junpei Shimamura, Ikumi Koga, Kento Hayata, Ryusuke Nibu, Yasuaki Koga, Hiroaki Kondo, and Hiroshi Furukawa

1 Introduction

Sensing equipment has decreased significantly in cost and has spread rapidly in this decade, even though its measurement accuracy and communication performance have improved. Consequently, wireless sensing systems have been widely studied and developed. Setting up a wireless sensor network (WSN) in a wide area requires installing large amounts of wireless communication equipment, viz., sensor gateways, and the installation costs tend to be high, depending on the extension of the sensing area.

Similar problems are occurring in setting up Wi-Fi networks. Currently, the importance of traffic off-loading from cellular networks to Wi-Fi networks is increasing owning to the rapid growth of network traffic [1] caused by the proliferation of smartphones and tablet devices. To extend Wi-Fi coverage areas, a significant number of Wi-Fi access points (APs) are being installed in urban areas. However, installation and maintenance costs become high, depending on the increase in the number of APs. To assist in alleviating this concern, we developed a

H. Togashi (✉) · H. Furukawa
Faculty of Information Science and Electrical Engineering, Kyushu University, Fukuoka, Japan
e-mail: togashi@mobcom.ait.kyushu-u.ac.jp

R. Abe · J. Shimamura · I. Koga · K. Hayata · R. Nibu · Y. Koga
Graduate School of Information Science and Electrical Engineering, Kyushu University, Fukuoka, Japan

H. Kondo
School of Engineering, Kyushu University, Fukuoka, Japan

© Springer Nature Switzerland AG 2020
Y. Liu et al. (eds.), *Smart Sensors and Systems*,
https://doi.org/10.1007/978-3-030-42234-9_6

wireless AP that uses wireless backhaul technology. We expect that this technology can alleviate the problems related to setting up wide wireless sensing areas. This article considers how wireless backhaul technology can be used in WSNs.

The remainder of this article is organized as follows. Section 2 briefly describes related technologies. Section 3 describes our basic sensing framework utilizing wireless backhaul technology. In Sect. 4, applications working in this framework are shown. Section 5 presents our concluding remarks.

2 Related Technologies

2.1 WSNs and Sensor and Actuator Networks (SANs)

A sensor network is a framework for observing the surrounding environment, e.g., temperature, illuminance, and noise, in a particular area. A typical sensor network consists of sensor nodes, sensor gateways, database servers, and access networks. Each sensor node is used to collect surrounding environmental data and transmit them to database servers. Sensor gateways are located in the same network as database servers and aggregate communication between sensor nodes and database servers.

WSNs are sensor networks that partially or fully use wireless communication technology. In some WSNs, sensor nodes can relay data transmitted from other nodes. Wi-Fi, Bluetooth, Bluetooth low energy, and LoRa technologies are often used in the infrastructures of WSNs. Wi-Fi is standardized as the IEEE 802.11 series [2] and is certified to have interoperability by the Wi-Fi alliance [3]. Most commoditized wireless access points (APs) use Wi-Fi technology. Bluetooth [4] is a technology for exchanging data over short distances (less than 10 m) and is standardized as IEEE 802.15.1 [5]. Bluetooth low energy (Bluetooth LE, BLE) is a type of Bluetooth technology that has low power consumption, data rate (in the range of 10 kbps), and latency (ms). LoRa (LoRaWAN) [6] is a wireless communication technology for an Internet of Things network and is standardized as IEEE 802.15.4 g. LoRa has a long communication distance (km) and low data rate (kbps).

A SAN is a combined framework consisting of a sensor network, data analysis functionalities, and actuators. Data analysis functionalities provide several kinds of analysis results, e.g., statistical calculation, forecasting, and situation recognition, on the basis of gathered sensor data. Actuators take appropriate actions with regard to the surrounding environment, e.g., lighting adjustment and airflow control, depending on the analysis results. Using machine learning algorithms in a SAN can provide more appropriate and natural actuation after a specific duration of continuous operation.

2.2 Wireless Backhaul Technology

Small cells are the key technology in setting up a wide Wi-Fi coverage area and achieving a large network capacity. To connect small cells to the core network (e.g., Internet), backhaul is essential. However, wired backhaul method is uneconomical and convenient, because all APs are wired to the core network. And achieving wide Wi-Fi coverage areas requires installing huge numbers of cables.

Wireless backhaul is a wireless multi-hop network in which APs are linked wirelessly with the capability of relaying packets. In a wireless backhaul network, a few APs (we refer to them as core nodes) are wired to the core network and serve as gateways connecting the wireless multi-hop network to the Internet. Therefore, wireless backhaul is advantageous in achieving a wide Wi-Fi coverage area, because this technology can reduce wiring cost significantly. For example, as in Fig. 1, the Wi-Fi coverage area can be set up by placing wireless APs with only one AP wired to the core network. However, mobile devices in this area can access the Internet, regardless of the AP with which a device is associated.

2.3 PCWL-0200

PCWL-0200 [7] is a wireless local area network (LAN) AP developed in our laboratory that uses wireless backhaul technology. This AP can set up a Wi-Fi coverage area without manually configuring frame relay routes among APs; the route is determined autonomously on the basis of the radio wave propagation environment, e.g., the received signal strength indications observed between each two APs. PCWL-0200 has been commercialized, and a significant number of Wi-Fi

Fig. 1 Introduction of wireless backhaul network. This technology enables extending a large Wi-Fi coverage area without laying huge numbers of access cables; only one AP is required to be wired to the core network

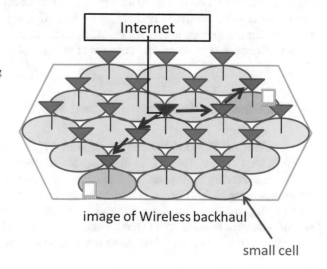

image of Wireless backhaul

small cell

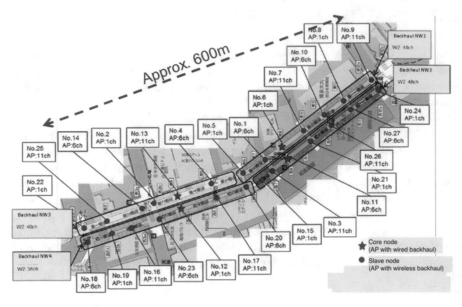

Fig. 2 Actual Wi-Fi access area set up using our Wi-Fi APs PCWL-0200; the length of the area is approximately 600 m, with only four APs wired to the core network

access areas set up using the APs are in practical use. Figure 2 shows an example of an approximately 600 m-long Wi-Fi access area using our Wi-Fi AP [8]. This Wi-Fi area is capable of accepting 4000 distinct users in a day, even though only four APs are wired to the core network.

Another feature of PCWL-0200 is that this AP can configure frame relay routes among the APs dynamically. By periodically reconfiguring a relay route, these APs can maintain a relay route with a specific level of communication quality and performance. This feature is suitable for mobile APs, e.g., APs located on moving vehicles, for comprising a wireless network, because this AP can reconfigure relay routes according to the locational changes of an AP.

3   Wireless Backhaul Network and WSNs

Our wireless AP PCWL-0200 can set up wide Wi-Fi access areas installing only a few access cables. This AP is also capable of connecting with a wired LAN device and serving as a gateway connecting the device to a wireless backhaul network. Owing to this feature, sensor gateways equipped with LAN connectors can transfer data to networked servers through a wireless backhaul network by wiring sensor gateways to slave APs simply. Note that connections between sensor nodes and gateways can be established using any optional technology. Data from sensor nodes are transmitted to a networked server via a wireless backhaul network; the AP referred to as the core serves as a gateway to connect this wireless network to the core network. Introducing wireless backhaul technology into WSNs leaves communications among sensor gateways unwired; this can reduce installation and maintenance costs of WSNs.

Figure 3 shows the components of an example WSN that uses wireless backhaul technology. PCWL-0200s are used to set up a wireless access infrastructure. Each sensor gateway is connected to the nearest adjacent slave AP and transmits data from sensor nodes to the networked server via the wireless backhaul network. Data communications among slave APs and between slave APs and the core AP are performed wirelessly; each slave AP is wired only to a sensor gateway with considerably fewer cables than in ordinary WSNs. In this framework, the core AP is wired to the core network, and sensor data are stored in the networked server. In the event that a WSN uses a local server, the core AP is wired only to the server.

The functionality of this WSN can be enhanced by connecting several types of servers to the database. For example, sensing results can be displayed visually using a web browser by connecting a web server to the database. To set up a SAN on this framework, a data analysis server and a control server for operating several types of actuators, e.g., cylinders and motors, are combined with this WSN. In the event that the system handles huge amounts of analysis data, high-performance servers and large-capacity databases for data analysis can be used in this sensing framework.

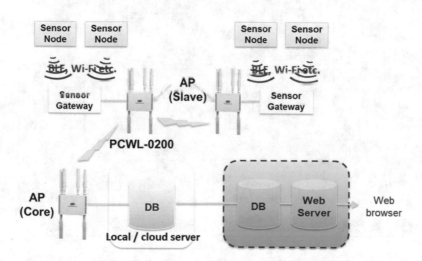

Fig. 3 Example WSN with wireless backhaul technology. Each sensor gateway is wired to the AP and relays sensing data gathered from surrounding sensor nodes to a server connected to the AP via a wireless backhaul network

3.1 Sensor Data Visualization Framework

We developed a sensor data visualization framework in this WSN framework. Figure 4 shows an example of the visualized temperature data in a laboratory room. In this figure, the respective circles indicate the locations of sensor nodes, and each square indicates the location of a sensor gateway. The sensing area is divided into a 0.2 m × 0.2 m grid, each unit of which is colored according to a sensor measurement as indicated in the legend shown at the bottom; higher values are indicated by red, and lower values are indicated by blue. This visualization framework uses a simple interpolation technique to estimate the environment of a particular sensing area from spatially sparse data. The environment of a particular sensing area can be recognized visually.

Our visualization framework can also show the time-series trend of each sensor node measurement. Figure 5 shows an example of a time-series plot. The horizontal axis corresponds to measuring times, and the vertical axis corresponds to measured values. Each dot is colored in the same manner as in the case of the room environment visualization.

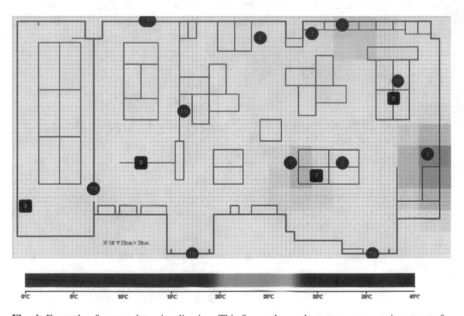

Fig. 4 Example of sensor data visualization. This figure shows the temperature environment of a particular room. Each circle indicates the location of a sensor node, and each square indicates the location of a sensor gateway

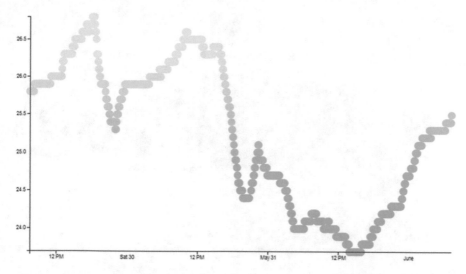

Fig. 5 Example of a time–series plot. Each dot is colored according to its measurement value

4 Applications of Wireless Backhaul Sensor Networks

This section introduces four applications using wireless backhaul sensor networks, a ubiquitous camera network, a Wi-Fi tag tracking system, a criminal fishing system, and a networked vehicle.

4.1 Ubiquitous Camera Network

A wireless camera network can be set up by connecting network cameras to slave APs. We refer to this framework as a ubiquitous camera network. Figure 6 shows the components of such a network. Each single-board computer with a universal serial bus (USB) camera is wired to a slave AP. Several types of information can be obtained by analyzing the camera images shot in the vicinity of the cameras and stored in the database.

However, transmission of many high-resolution images consumes sufficient wireless network bandwidth to degrade communication quality, e.g., to increase packet loss and network delay. To alleviate this problem, we considered analyzing camera images using single-board computers and transmitting only the analysis results. TensorFlow [9] is used in the ubiquitous camera network to analyze camera images. Inception-v3 [10], an image recognition model that works on TensorFlow, can recognize what is photographed in a particular picture and enumerate it as textual information. Recognition results (we refer to them as spatial textual information) are shown as five candidate synsets registered in WordNet [11] with

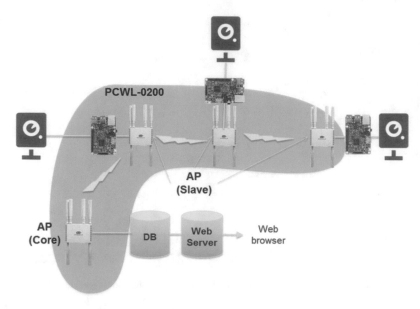

Fig. 6 Example of a ubiquitous camera network. Each single-board computer with a USB camera is wired to a slave AP. Spatial textual information is transmitted to a server connected to the AP referred to as the core via wireless backhaul network

Fig. 7 Applications of spatial textual information. Left: vacancy of a parking area can be recognized in an image by printing logos on each parking space. Right: vehicle congestion can be recognized by printing logos on vehicle floors

a confidential score on each synset. In the ubiquitous camera network, each single-board computer executes TensorFlow and transmits spatial textual information to the database server. Generally, the data size of a still image is measured in megabytes, and spatial textual information measures less than 1 KB.

Through the use of spatial textual information, occupancy and vacancy of a particular region can be recognized by observing the region visually. For example, in Fig. 7, the vacancy of a parking area can be recognized by printing logos, i.e., pictures easy to distinguish from each other, on each parking spot and monitoring the area. When the logo can be recognized, the parking spot is vacant. In the same manner, congestion on each train vehicle can be observed by printing logos on the floor and observing the floor through captured images; when many logos can be recognized, the vehicle is less crowded.

4.2   Wi-Fi Tag Tracking System

Demand for care services, especially for children and the elderly, has been increasing. Several Global Positioning System (GPS)-based pedestrian tracking systems are being investigated and developed [12, 13], but GPS cannot provide accurate position estimates in high-rise areas and indoor environments. To alleviate this issue, we propose a pedestrian tracking system that uses Wi-Fi beacons held by target persons and Wi-Fi APs placed widely and densely in a specified area. The proposed tag tracking system uses probe request signals broadcast by mobile devices to estimate their positions. Our pedestrian tracking system uses Wi-Fi APs to set up a positioning area and Wi-Fi beacons that broadcast probe request signals periodically. Each beacon's position, i.e., target user's position, in this area is estimated using a networked server, on the basis of the probe request signals broadcast by the beacons and captured by the APs. The position estimates are stored in a networked database. The trajectories of each beacon are estimated on the basis of the position estimates, taking the floor plan of the area into account.

The system components are shown in Fig. 8. Each target user holds a Wi-Fi beacon. PCWL-0200s are used to set up a Wi-Fi coverage area, i.e., a positioning area, and capture the probe request signals broadcast by surrounding beacons. These APs transmit these signals to a positioning server implemented in Python via the wireless backhaul network at the request of the server. A single probe request signal can be captured on multiple APs. Each beacon's position is estimated on the positioning server through analysis of these signals. Position estimates are stored in a MongoDB [14] positioning database. The trajectory of each beacon is estimated from its position estimates and stored on the positioning server. These trajectories are displayed visually using the D3.js [15] JavaScript library. The web server is implemented using Django [16].

Figure 9 shows a screenshot of the proposed pedestrian tracking system. The main panel (Panel 1) displays the estimated trajectory of each Wi-Fi beacon. The filled numbered circles indicate the locations of the respective APs, and the respective arrows and unfilled circles, colored as shown in the legend, indicate estimated trajectories of individual beacons. The arrows indicate the route each beacon has traveled for a specific duration, and the unfilled circles indicate locations at which the beacon has remained. In Fig. 9, Beacon 1 (blue) has remained near AP12, and Beacon 3 (green) has traveled from AP7 to AP5. This panel also has functionalities to show the floor on which each beacon is located and to change the

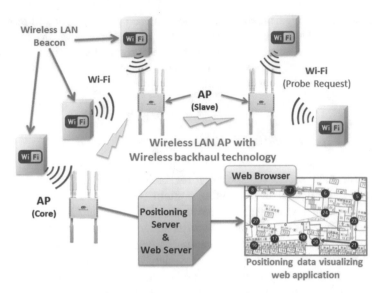

Fig. 8 Components of the tag tracking system. Each beacon's position is estimated on the positioning server by analyzing the probe request signals captured by wireless APs. Probe request signals are gathered to the server via the wireless backhaul network

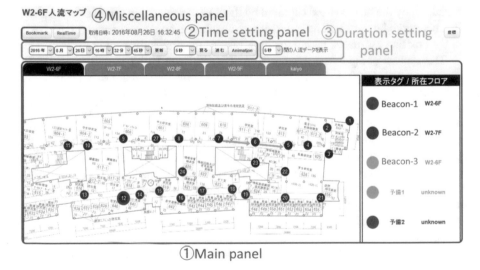

Fig. 9 Screenshot of the pedestrian tracking system. The arrows and unfilled circles colored as shown in the legend indicate the estimated positions of respective beacons

floor to display estimated trajectories. The floor on which each device is located is displayed in the right-hand table of the main panel. This table also indicates the names of users who hold Wi-Fi beacons.

4.3 Criminal Fishing System

Criminal activities, such as graffiti, shoplifting, larceny, and kidnapping, occur in urban areas, causing various types of social damage. We expect that utilizing large numbers of APs in urban areas as a component of a criminal investigation assistance system can reduce the frequency of criminal activity and improve public safety. The proposed "criminal fishing system" uses a large number of wireless LAN APs and cameras. This system enumerates candidate media access control (MAC) addresses of a culprit's mobile device from probe request signals gathered by APs during the period in which a culprit remains near the incident scene.

Figure 10 illustrates the components of the system. The principal components are an administrative server, PCWL-0200s (Wi-Fi APs), and USB cameras. Each AP gathers probe request signals and periodically transmits radio wave fingerprints to the administrative server. In this system, we refer to a probe request signal as a radio wave fingerprint, because it contains a MAC address basically unique to a particular network interface. Each AP executes motion, an image-capturing program that captures images when the program recognizes moving objects. These images are transmitted to the administrative server via a wireless backhaul network. The transmitted radio wave fingerprints and images are stored in the database on the server. The administrative server is equipped with a web application for enumerating the candidate MAC addresses of a culprit's device from a huge amount of stored

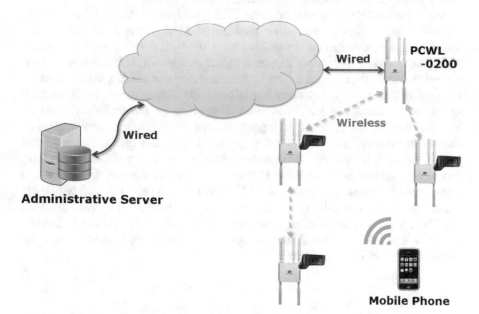

Fig. 10 Criminal fishing system. Each access point gathers probe request signals broadcast by surrounding devices. These signals and photos near each AP are transmitted to the administrative server via a wireless backhaul network

Fig. 11 Example of surveillance assistance using candidate MAC addresses. After candidate MAC addresses are enumerated, the system monitors the appearance of the MAC address. When the culprit's MAC address is observed in one of the investigation areas, the system issues an alert and captures photos around the area

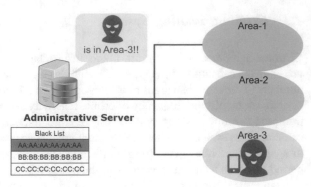

data. A facility administrator uses the web application to determine the duration that the culprit remains near the scene of the incident while checking the stored images. The administrative server enumerates candidate MAC addresses during this time.

Even if the detailed appearance of the culprit cannot be obtained near the scene of the incident, it can be obtained later at another location by surveilling the enumerated MAC addresses, as shown in Fig. 11. When a MAC address is captured in a particular area, the appearance of the device owner can be obtained from the camera image shot near there. The probability of identifying the culprit can be improved by locating APs over wide areas, with high density.

We must consider how our criminal fishing system works if the culprit has a mobile device with a spoofed or randomized MAC address. In short, MAC spoofing and randomization themselves have no critical effect on this system. These technologies sometimes trigger frequent changes of MAC address, which undeniably affect the criminal fishing system. In the case in which a MAC address is changed very frequently, e.g., every few minutes, the address cannot be enumerated as one of the candidate MAC addresses. Depending on the frequency of MAC address changes, the criminal fishing system can assist criminal investigation as follows. If the MAC address is not changed for several days (perhaps even several hours), the usual appearance of the owner of the device, i.e., the candidate culprit, can be obtained by monitoring the MAC address. In this manner, the system can assist crime investigations, even if the MAC address has been changed before the candidate culprit is actually questioned regarding the incident. In a case in which the MAC address is rarely changed regardless of whether the address is factory-assigned or spoofed, this MAC address itself can be considered as evidence that the person is a culprit in the incident.

4.4 Networked Vehicle and WSN

As mentioned in Sect. 2.3, a PCWL-0200 AP is capable of configuring frame relay routes among the APs comprising a wireless network autonomously and dynamically. This feature enables a WSN to use networked vehicles equipped with AP functionality and several types of sensors and actuators.

A prototype of the networked vehicle, PicoRover, has been developed based on this concept. The vehicle is shown in Fig. 12. Its principle components are a PCWL-0200 Wi-Fi AP, a Raspberry Pi, a USB camera, and a motor driver integrated circuit. To operate PicoRover, a remote computer sends control commands to the vehicle via a wireless backhaul network. The Raspberry Pi decodes the commands and operates the vehicle using a motor driver, sending camera images to a remote computer via the wireless backhaul network at the user's request.

PicoRover can be enhanced to a networked sensing vehicle by attaching several types of sensors to a Raspberry Pi on the vehicle. Sensing data can be transmitted to a networked server via a wireless backhaul network in the same manner as camera images. Using networked sensing vehicles in a WSN can reduce the number of sensor nodes significantly. The total installation cost of sensor nodes can be reduced, even though the installation cost of a network sensing vehicle is estimated to be higher than that of ordinary sensor nodes. Maintenance costs of sensor nodes can also be reduced by introducing networked sensing vehicles, since the total number of sensor nodes is reduced. Another feature of such a vehicle is that the sensing area of a WSN can be formed dynamically and flexibly using these vehicles, since the vehicles are capable of configuring frame relay routes among themselves autonomously and dynamically. For example, in Fig. 13, the sensing area is formed by driving each sensing vehicle remotely, owing to the functionality of the PCWL-

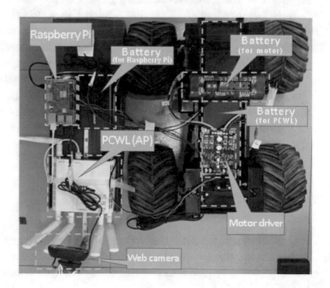

Fig. 12 PicoRover. This vehicle is equipped with a PCWL-0200 and works as a moving AP. Control commands are sent to the vehicle via a wireless backhaul network. A Raspberry Pi decodes the commands and operates the vehicle, which is equipped with a USB camera whose images are transmitted to a remote computer

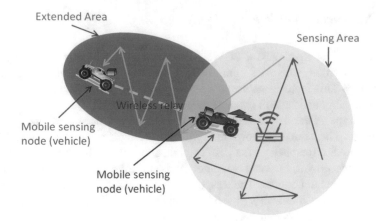

Fig. 13 Extension of sensing area using networked sensing vehicles. The sensing area can be extended and adjusted by driving these vehicles. Frame relay routes among sensing vehicles are configured autonomously and dynamically, and control commands and sensing data are exchanged using a wireless backhaul network

0200 AP. In this manner, the sensing area can be extended to locations that ordinary WSNs find difficult to cover, e.g., where access cables are difficult to lay, where sensor gateways and nodes cannot be placed permanently, or where persons have difficulty entering the region.

To measure the environment of a particular area continuously and autonomously, a sensing vehicle must be equipped with the functionality of inspecting the area and measuring the surrounding environment autonomously. The measurement frequency at a location, i.e., the frequency of passing vehicles, can be adjusted by changing the number of sensing vehicles and their inspection routes.

Introducing networked sensing vehicles in a WSN also enables on-demand sensing. Users of this sensing framework can obtain environmental information anytime and anywhere at the request of the user. In an ordinary WSN, it is difficult to satisfy this kind of demand; measuring the environment of a specific location requires that a person go there and place sensors in advance. Using networked sensing vehicles in WSNs can eliminate this preparation. Switching between autonomous and on-demand sensing can be expected to increase the usability of the vehicle. Coexistence of multiple sensing methods will be a unique characteristic of a WSN with networked sensing vehicles.

5   Conclusion

This paper described the introduction of wireless backhaul technology into WSNs. Doing this unwires communication among sensor gateways. Because communication between sensor nodes and gateways can be established with any optional technology, it is easy to replace the network infrastructure of existing WSNs. Installation and maintenance costs of WSNs can be reduced by using this technology, since the number of access cables is reduced significantly.

Four application systems based on this sensing framework were described, a ubiquitous camera network, a Wi-Fi tag tracking system, a criminal fishing system, and a networked vehicle. Using networked vehicles in a sensor network can reduce the number of sensor nodes significantly. The total installation and maintenance cost of sensor nodes can be reduced, even though the installation cost of a networked sensing vehicle is estimated to be much higher than that of an ordinary sensor node. Moreover, using a networked sensing vehicle is also beneficial because a user of this sensing framework can measure environmental information anytime and anywhere at the request of the user.

This sensing framework has the potential to be used with many types of sensing systems, because any kind of sensing equipment can be used with it. We would like to conduct further investigations aiming at the discovery of further application scenarios, and we expect to report novel sensing applications on other occasions.

References

1. Ericsson Mobility Report–Mobile World Congress edition, February 2016, [Online]. http://www.ericsson.com/res/docs/2016/mobility-report/ericsson-mobility-report-feb-2016-interim.pdf
2. IEEE 802.11, The Working Group for WLAN Standards, [Online]. http://www.ieee802.org/11/
3. Wi-Fi alliance, [Online]. https://www.wi-fi.org/
4. Bluetooth SIG, Inc., Bluetooth Technology Website, [Online]. https://www.bluetooth.com
5. IEEE 802.15 Working Group for Wireless Specialty Networks (WSN), [Online]. http://www.ieee802.org/15/
6. LoRa Alliance, [Online]. https://www.lora-alliance.org/
7. PicoCELA Inc., PCWL-0200 [Online]. http://jp.picocela.com/12477.html, (in Japanese)
8. Tenjinchikagai, Japan's largest Wi-Fi street "Tenjin underground shopping center", [Online]. http://tenchika.com/about/wifi/
9. TensorFlow, [Online]. https://www.tensorflow.org/
10. Szegedy C, et al., Rethinking the inception architecture for computer vision, [Online]. http://arxiv.org/abs/1512.00567
11. Princeton University, WordNet, [Online]. https://wordnet.princeton.edu/
12. Location Based Technologies, Inc., PocketFinder 3G global GPS+ trackers, [Online]. http://pocketfinder.com
13. Best 3G GPS tracker devices—Trackimo, [Online]. https://trackimo.com/
14. MongoDB Inc., MongoDB for GIANT ideas, [Online]. https://www.mongodb.org/
15. D3.js–Data driven documents, [Online]. https://d3js.org/
16. Django Software Foundation, Django: the web framework for perfectionists with dead-lines, [Online]. https://www.djangoproject.com/

Basics and Challenges in Acoustic Vehicle Sensing Using Sidewalk Microphones

Shigemi Ishida, Shigeaki Tagashira, and Akira Fukuda

1 Introduction

The past decade has seen the rapid development of ITS (intelligent transportation system). The main purpose of the ITS is to improve the safety, efficiency, dependability, and cost effectiveness of transportation systems. Products such as car navigators make the ITS more prevalent nowadays.

In the ITS, vehicle detection is one of the fundamental tasks. Automatic vehicle sensing systems have been widely deployed to detect vehicles. The deployment of the vehicle sensing system is, however, limited to high traffic roads because of high deployment and maintenance costs of the system in terms of roadwork closing a target road section. Current automatic vehicle sensing systems also suffer from a motorbike detection problem because of small coverage of vehicle sensors.

We therefore developed a simple vehicle sensing system using acoustic sensors, namely, microphones. Stereo microphones are installed at a roadside and capture acoustic signals generated from vehicle tires to detect passing vehicles. We can detect vehicles on multiple lanes from one side of a road because sound signals are diffracted over obstacles, which drastically reduces roadwork costs for deployment and maintenance.

Some literature studies on a vehicle sensing system using acoustic sensors [1, 3–5]. Vehicle sensing systems presented in these studies rely on a microphone array to draw a *sound map*, i.e., a map of time difference of vehicle sound on different

S. Ishida (✉) · A. Fukuda
ISEE, Kyushu University, Fukuoka, Japan
e-mail: ishida@f.ait.kyushu-u.ac.jp; fukuda@f.ait.kyushu-u.ac.jp

S. Tagashira
Faculty of Informatics, Kansai University, Osaka, Japan
e-mail: shige@res.kutc.kansai-u.ac.jp

© Springer Nature Switzerland AG 2020
Y. Liu et al. (eds.), *Smart Sensors and Systems*,
https://doi.org/10.1007/978-3-030-42234-9_7

microphones. The studies manually analyzed the sound map and demonstrated the feasibility of vehicle sensing using the sound map.

We extended these sound map studies and developed an automatic vehicle sensing system. We have experimentally demonstrated that our vehicle sensing system had successfully detected vehicles with an F-measure of 0.92 using a state-machine based vehicle detection algorithm [7] or template matching [6].

In this paper, we review the basics of our acoustic vehicle sensing system including a theoretical background of the sound mapping and initial design as well as implementation of the system. We also present challenges we are working for accuracy improvement at multi-lane high traffic roads.

The remainder of this paper is organized as follows. Section 2 overviews related works on vehicle sensing. Section 3 describes our vehicle sensing system including a theoretical background and extends the system in Sect. 4 to address practical issues. In Sect. 5, we implement our vehicle sensing system and conduct experiments to evaluate the detection performance. Section 6 indicates challenges we are facing, implying future research directions. Finally, Sect. 7 summarizes the paper.

2   Related Works

Vehicle sensors are divided into two types: intrusive and non-intrusive.

Loop coils and photoelectric tubes are categorized into the intrusive vehicle sensors. These vehicle sensors need to be installed under the road surface, which results in high costs due to roadwork closing a target road section. Loop coils and photoelectric tubes also suffer from a motorbike detection problem; motorbikes are highly missed because of small sensor coverage.

The non-intrusive sensors are based on laser, infrared, ultrasound, radar, or camera. The non-intrusive vehicle sensors are installed above or by a road for better performance. Deployment above a road requires high installation and maintenance costs in terms of roadwork. Although installation of roadside non-intrusive vehicle sensors requires no roadwork, the roadside sensors can be applied to single lane roads. Most of non-intrusive sensors are based on laser, infrared, or ultrasound, which have small coverage suffering from the motorbike detection problem.

To reduce installation and maintenance costs, camera-based vehicle sensors using CCTVs installed in the environment have been proposed [2, 9]. CCTVs, however, are available in limited areas, especially in city areas. Camera location and angle are designed for security surveillance not for vehicle sensing, resulting in low detection accuracy in bad weather conditions.

On the contrary, acoustic approach is a promising candidate for vehicle sensing at a low installation and maintenance costs. Using stereo microphones at a sidewalk, we can locate a sound source, i.e., a vehicle on a road. Vehicles on multiple lanes are detected from one side of a road because vehicle sounds are diffracted over other vehicles.

Several studies have reported on a vehicle monitoring system using acoustic sensors. Forren et al. and Chen et al. proposed traffic monitoring schemes using a microphone array [3–5]. The monitoring schemes draw a sound map, i.e., a map of time difference of vehicle sound on different microphones and analyze the sound map to monitor traffic. The monitoring schemes are missing design details of vehicle detection from the sound map. Microphone array is installed in a high height configuration at a roadside to monitor vehicles on multiple lanes. The high height configuration results in high installation and maintenance costs in terms of safety installation and maintenance.

Barbagli et al. reported an acoustic sensor network for traffic monitoring [1]. The acoustic sensor network installs sensor nodes at roadsides. Each sensor node draws a sound map and combines the sound map with an energy detection result to monitor traffic flow distribution. The sensor network requires many sensor nodes at both sides of the road to monitor real-time traffic flow, resulting in high deployment and maintenance costs. The paper also lacks an evaluation of vehicle detection accuracy because main focus of the paper is on traffic flow monitoring with small energy consumption.

3   Acoustic Vehicle Sensing System

3.1   System Overview

Figure 1 depicts the overview of our acoustic vehicle sensing system. Our vehicle sensing system consists of three components: a sound retriever, sound mapper, and vehicle detector. A sound retriever is two microphones followed by LPFs (low-pass filters) reducing high frequency environmental noise. Stereo microphones are installed at a sidewalk of a road and capture acoustic signals generated by vehicle tires. The sound mapper calculates cross-correlation between sounds on the stereo microphones to estimate TDOA (time difference of arrival) and generate a sound map. A vehicle detector finally detects vehicles using a state-machine based algorithm.

Following subsections describe each component.

Fig. 1 Overview of acoustic vehicle sensing system

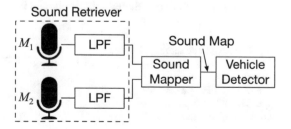

3.2 Sound Retriever

Figure 2 depicts the microphone setup. Two microphones M_1 and M_2 are installed parallel to a road at a distance of L. The two microphones are separated by D. Sound signals generated by vehicle tires travel to the microphones with different distances d_1 and d_2, therefore form a time difference between signals on M_1 and M_2. Let x be the location of a vehicle. When a vehicle on a road is far from the microphones, difference $|d_1 - d_2|$ of sound traveling distance becomes maximum at $|d_1 - d_2| \simeq D$. The maximum time difference Δt_{max} of sound arrival on the microphones is therefore calculated as

$$|\Delta t_{max}| = \frac{D}{c}, \tag{1}$$

where c is the speed of sound in air. As a vehicle moves from left to right in the figure, the time difference Δt increases from $-\Delta t_{max}$ to Δt_{max}.

To reduce the influence of environmental noise, we apply a LPF (low-pass filter). The majority of frequency components of sound signals generated by vehicle tires is under 2.0 kHz [11]. The cut-off frequency of the LPF is therefore set to 2.5 kHz including a margin. Because tires generate sound signals for all types of vehicles, our vehicle sensing system is capable of detection of all types of vehicles.

3.3 Sound Mapper

A sound map is time difference of signals on the two microphones as a function of time. Sound traveling distances d_1 and d_2 in Fig. 2 are calculated as

$$d_1 = \sqrt{\left(x + \frac{D}{2}\right)^2 + L^2}, \tag{2}$$

$$d_2 = \sqrt{\left(x - \frac{D}{2}\right)^2 + L^2}. \tag{3}$$

Sound delay Δt between the two microphones is therefore

Fig. 2 Microphone setup. Two microphones M_1 and M_2 are installed at a roadside parallel to a road at a distance of L. The microphones are separated by distance of D

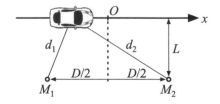

$$\Delta t = \frac{d_1 - d_2}{c}$$

$$= \frac{1}{c} \left\{ \sqrt{\left(x + \frac{D}{2}\right)^2 + L^2} - \sqrt{\left(x - \frac{D}{2}\right)^2 + L^2} \right\}. \tag{4}$$

Equation (4) indicates that the location of a vehicle can be calculated from sound delay Δt. The sound delay is estimated using a cross-correlation function defined as

$$R(t) = s_1(t) * s_2(t), \tag{5}$$

where $s_1(t)$ and $s_2(t)$ are signals received by the two microphones and $*$ denotes the convolution operation.

The cross-correlation function $R(t)$ becomes maximum at $t = \Delta t$ when the microphones receive the same sound signals with time shifted by Δt, i.e., $s_1(t) = s_2(t + \Delta t)$. We can estimate sound delay Δt by finding a peak of $R(t)$.

We use a GCC (generalized cross-correlation) function [8], which is commonly used in acoustic source localization, instead of a normal cross-correlation function to increase robustness against environmental noise. The sound mapper divides sound signals into chunks with a small window and applies the GCC to the each chunk to estimate a sound delay. Plotting the sound delay as a function of time gives a sound map.

Figure 3 shows a typical sound map. As a vehicle passes in front of the microphones, Δt rises up or drops down on the sound map drawing an S-curve; direction of the S-curve corresponds to the direction of the vehicle.

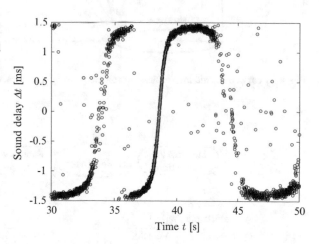

Fig. 3 Example of sound map. Vehicle passing is drawn as an S-curve on sound map. Direction of the S-curve depends on the direction of the vehicle passing

3.4   Vehicle Detector

Figure 4 illustrates an S-curve on a sound map when a vehicle passes in front of microphones from left to right. We can divide an S-curve into three sub-curves as shown in Fig. 4; sub-curves 1, 2, and 3 are observed when a vehicle coming toward microphones, passing in front of microphones, and going away from microphones, respectively. The sub-curves 1 and 3 are close to asymptotes $\Delta t = \pm \Delta t_{max}$ given by Eq. (1).

Order of sub-curves depends on vehicle direction because the direction of an S-curve depends on vehicle passing direction. We separately apply a vehicle detection algorithm for each vehicle direction on a sound map. This subsection describes vehicle detection algorithm using an S-curve of a vehicle passing from left to right, as an example. The detection algorithm is a state machine that tracks sub-curves on a sound map.

Figure 5 illustrates a state machine diagram of a vehicle detection process. The detection process consists of four states. The process starts from a *sub-curve 1*

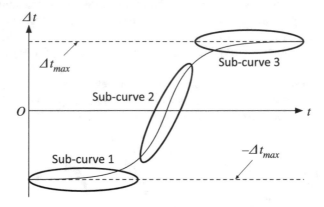

Fig. 4 S-curve on sound map indicating vehicle passing. The S-curve consists of three sub-curves

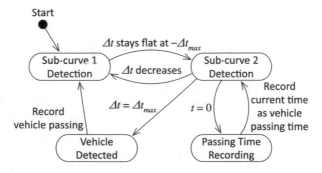

Fig. 5 State machine diagram of vehicle detection process (for vehicles passing from left to right). State transitions are based on sound delay Δt

detection state. Sub-curve 1 is close to an asymptote $\Delta t = -\Delta_{max}$. If the sound delay Δt stays flat approximately at $\Delta t \simeq -\Delta t_{max}$ for the specific duration, the detection process switches to a *sub-curve 2 detection* state.

In the sub-curve 2 detection state, the detection process tracks the sound delay Δt increasing as time increases. When the sound delay $\Delta t = 0$, the detection process temporarily switches to a *passing time recording* state to make a tentative record of vehicle passing time.

When Δt reaches Δt_{max}, the detection process switches its state to a *vehicle detected* state and goes back to the sub-curve 1 detection state to start a next detection process. In a sub-curve 2 detection state, the detection process returns to a sub-curve 1 detection state when Δt decreases to restart the detection process.

4   Practical Issue

4.1   Big Vehicle Detection Problem

Sound delay draws an S-curve for a vehicle, as shown in Fig. 3. For big vehicles such as buses, trucks, sound delay partially splits into two curves, resulting in failure of vehicle detection.

Figure 6 shows an example of sound map when a bus is passing in front of microphones. An S-curve partially splits into two curves around $\Delta t = 6$ s. The maximum separation between the two curves is approximately 1 ms. A curve on a sound map indicates a vehicle; two curves mistakenly indicate two vehicles and result in a false positive detection.

The two curves on a sound map are generated by front and rear tires. Consider the case of a bus passing right in front of microphones as shown in Fig. 7. Let l be a wheelbase of the bus. Sound delays Δt_F and Δt_R of front and rear tires are

Fig. 6 Example of sound map when bus is passing

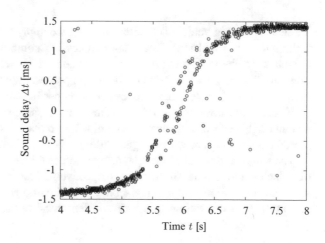

Fig. 7 Bus passing right in front of microphones. Delays Δt_F, Δt_R of sound signals generated from front and rear tires are significantly different because of long wheelbase

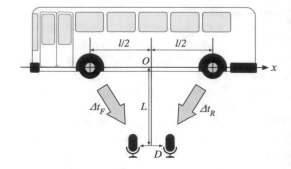

calculated by Eq. (4). We derive difference δ between the sound delays of front and rear tires as

$$\delta = \Delta t_R - \Delta t_F, \tag{6}$$

$$\Delta t_F = \frac{1}{c}\left\{\sqrt{\left(\frac{-l+D}{2}\right)^2 + L^2} - \sqrt{\left(\frac{-l-D}{2}\right)^2 + L^2}\right\}, \tag{7}$$

$$\Delta t_R = \frac{1}{c}\left\{\sqrt{\left(\frac{l+D}{2}\right)^2 + L^2} - \sqrt{\left(\frac{l-D}{2}\right)^2 + L^2}\right\}. \tag{8}$$

When $l = 4.5$ m, $D = 0.5$ m, and $L = 4$ m, the sound-delay difference δ is calculated to be 1.4 ms. The sound-delay difference is considerable compared to the maximum sound delay $\Delta t_{max} \approx 1.5$ ms in Fig. 3.

4.2 Image Processing for Sound Map

To address the big vehicle detection problem, we apply a simple image processing technique to a sound map. The image processing combines separated two curves to generate a single bold curve. No modification on the vehicle detection algorithm presented in Sect. 3.4 is required. The image processing also reduces the influence of environmental noise and improves detection performance.

Figure 8 illustrates an overview of image processing. We replace each point on a sound map with a translucent rectangle. The rectangles overlap each other and reproduce a bold curve with a vague outline.

The height and width of the rectangles are determined from the maximum length of vehicles and a road speed limit, respectively. The height of the rectangles is set to the maximum sound-delay separation. Maximum sound-delay separation is calculated by substituting the maximum length of vehicles for l in Eq. (6). The

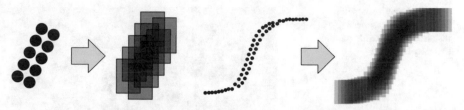

Fig. 8 Overview of image processing. Each point on a sound map replaced with a translucent rectangle. The rectangles overlap and generate a bold line for S-curves that indicate vehicle passing

Fig. 9 Sound map after image processing derived from Fig. 6

width of the rectangles should be set to a time duration such that a vehicle at the speed limit moves negligible distance compared to the vehicle length.

We apply our image processing technique to a sound map shown in Fig. 6 and derive Fig. 9. Comparing the figures reveals the effect of the image processing; we can observe a single bold curve on a sound map for a vehicle.

5 Evaluation

5.1 Experiment Setup

We conducted experiments in our university evaluating the basic performance of our vehicle sensing system. Figure 10 shows an experiment setup. A target road has two lanes, one lane for each direction. Two microphones were installed at a sidewalk of the road. We recorded vehicle sound for approximately 30 min using a Sony PCM-D100 recorder with OLYMPUS ME30W microphones. The sound was recorded at a sampling frequency of 48 kHz and word length of 16 bits. We recorded video monitoring the target road, which was used as ground truth data.

The two microphones were separated by $D = 50$ cm, which was determined based on preliminary experiment results. In a preliminary experiment, we changed microphone separation from 50 to 150 cm and performed vehicle detection for approximately 20 min to compare vehicle detection performance. The distance L between road center and microphones was 2 m, which was physically restricted.

Fig. 10 Experiment setup.
Two microphones were
installed at a roadside parallel
to a two-lane road

Comparing the results derived by our system with the video, we evaluated the numbers of true positives (TPs), false negatives (FNs), and false positives (FPs), TP, FN, and FP are defined as the case that a vehicle detected when a vehicle passing, no vehicle detected when a vehicle passing, and a vehicle detected when no vehicle passing, respectively. We excluded true negatives (TNs), which is defined as the case that no vehicle detected when no vehicle passing, because TNs were not countable in our experiments. Using the numbers of TPs, FNs, and FPs, we also evaluated a precision, recall, and F-measure defined as follows:

$$\text{Precision} = \frac{\text{TP}}{\text{TP} + \text{FP}} \tag{9}$$

$$\text{Recall} = \frac{\text{TP}}{\text{TP} + \text{FN}} \tag{10}$$

$$\text{F}_{\text{measure}} = \frac{2 \cdot \text{Precision} \cdot \text{Recall}}{\text{Precision} + \text{Recall}}. \tag{11}$$

5.2 Experiment Results

Table 1 summarizes the numbers of TPs, FNs, and FPs. As described in Sect. 3.4, our vehicle sensing system separately detects vehicles for each vehicle direction. We therefore derived the numbers of TPs, FNs, and FPs for each vehicle direction and summed the results to retrieve a total result. A precision, recall, F-measure are calculated from the numbers of TPs, FNs, and FPs.

From Table 1, we can confirm that our vehicle sensing system successfully detected vehicles with an F-measure of 0.92. Our vehicle sensing system achieved the performance at the same level as existing sensors; the accuracy of a vehicle sensing system using a magnetic sensor was 95 % as reported in [10].

Table 1 Experiment results

	Left to right	Right to left	Total
TP	63	87	150
FN	11	15	26
FP	0	0	0
Precision	1.00	1.00	1.00
Recall	0.85	0.85	0.85
F-measure	0.92	0.92	0.92

Table 2 Experiment results vs. vehicle types

	Normal cars	Buses, trucks	Small cars	Motorbikes	All
TP	62	42	29	17	150
FN	17	4	1	4	26
FP	0	0	0	0	0
Precision	1.00	1.00	1.00	1.00	1.00
Recall	0.78	0.91	0.97	0.81	0.85
F-measure	0.88	0.95	0.98	0.89	0.92

The number of FPs was zero. Our vehicle sensing system exhibited high tolerance to environmental noise such as wind and people chattering. There was no mistake on detection of vehicle direction. We observed the same performance for both vehicle directions although left-to-right vehicles passed on the other side of the road.

Table 2 shows the experiment results for each type of vehicles. We confirmed that all types of vehicles were successfully detected with a minimum F-measure of 0.88. Our system exhibited the smallest F-measure for normal cars. This was mainly caused that there were many normal cars simultaneously passing in front of microphones. The numbers of simultaneous passing of normal cars, buses/trucks, small cars, and motorbikes were 27, 9, 8, and 6, respectively.

6 Challenges

To accurately detect vehicles at multi-lane high traffic roads, our vehicle sensing system faces three big challenges.

6.1 Sparse Sound Map

In our vehicle sensing system, we fully rely on a sound map to detect vehicles. When a sound map is sparse and indistinct, our vehicle detection algorithm fails to detect vehicles. A sound map tends to be sparse when multiple vehicles are in front of microphones.

Figure 11 shows an example of sparse sound map. When two vehicles are simultaneously passing in front of microphones in opposite directions, a sound mapper highly fails to estimate sound delay of the vehicle sound. A cross-correlation function of mixed sounds from two vehicles gives two peaks corresponding to the two vehicles. The peaks, however, exhibit lower value compared to a peak derived when one vehicle is passing because sound of one vehicle works as a noise to the other vehicle sound. The weakened sound signals have higher chance to be affected by environmental noise, resulting in a sparse sound map. Sparse sound map is more problematic at multi-lane high traffic roads because simultaneous passing occurs at a higher probability.

Figure 12 shows a GCC result when two vehicles are simultaneously passing in front of microphones. The GCC gives six peaks, the two of them at sound delay of −1.4 and 1.4 ms correspond to the two vehicles. The peak at −0.75 ms is caused by an environmental noise, which is greater than the peaks by the vehicles.

Fig. 11 Sparse sound map derived when two vehicles simultaneously passing in opposite directions. Sound map becomes sparse when two vehicles are simultaneously passing because a single value of sound delay is calculated at each time point

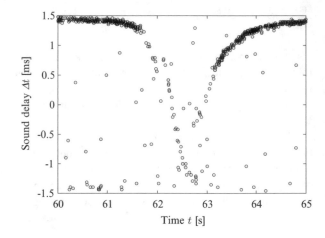

Fig. 12 GCC results when two vehicles are simultaneously passing in front of microphones. Peaks at around −1.4 and 1.4 ms correspond to two vehicles. The peak at −0.75 ms is caused by an environmental noise, which is the biggest peak in this GCC result

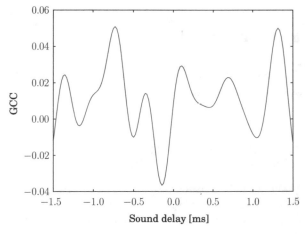

6.2 High Power Consumption

As described in Sect. 3.3, our vehicle sensing system relies on GCC (generalized cross-correlation) in a sound mapper. The GCC computation requires convolution computations for many times. Real-time computation of GCC therefore requires a computer with significant resources, resulting in high power consumption.

High power computers prevent large deployment because of restrictions on a power line. Although detected sensor data can be sent via wireless links, we still need a power cable. Energy saving is therefore an important aspect for battery operation.

As a suggestion, a two-step wake-up scheme might be employed in our vehicle sensing system. Ultra low-power vehicle detector using a non-accurate sensor roughly detects vehicles to wake the vehicle sensing system presented in this paper to accurately detect vehicles. Toward this goal, we need ultra low-power vehicle detectors.

6.3 Multiple Vehicle Detection Algorithm

The vehicle detection algorithm described in Sect. 3.4 is a simple but effective state machine that keeps track of S-curves on a sound map. The algorithm is, however, too simple to detect multiple vehicles at a time. The presented algorithm follows the increase or decrease of S-curves; S-curves lose their S-shape when S-curves are overlapping each other, resulting in detection failures even if we use a clear sound map.

Acoustic vehicle sensing system is capable of vehicle detections on multi-lane roads from one side of a road because sound signals are diffracted over obstacles. Toward robust vehicle detection at multi-lane high traffic roads, we need a new approach of sound map analysis. There are many works on pattern recognition algorithms. We believe that some of them can be useful in our vehicle sensing system.

7 Summary

In this paper, we presented an acoustic vehicle sensing system that comes with a low deployment cost. Our vehicle sensing system only relies on stereo microphones installed at a sidewalk of a road. We draw and analyze a sound map, i.e., a time-difference map of vehicle sound on the two microphones, to find S-curves that indicate vehicles. To reduce the false detections caused by big vehicles, we apply a simple image processing to a sound map generating bold S-curves. We conducted experimental evaluations and demonstrated that our vehicle sensing system accurately detected vehicles with an F-measure of 0.92. We also presented challenges at multi-lane high traffic roads, implying the future research directions.

Acknowledgements This work was supported in part by JSPS KAKENHI Grant Numbers JP15H05708, JP16K16048, JP17H01741 and the Cooperative Research Project of the Research Institute of Electrical Communication, Tohoku University.

References

1. Barbagli B, Manes G, Facchini R, Manes A (2012) Acoustic sensor network for vehicle traffic monitoring. In: Proc. IEEE int. conf. on advances in vehicular systems (VEHICULAR), pp 1–6
2. Buch N, Cracknell M, Orwell J, Velastin SA (2009) Vehicle localisation and classification in urban CCTV streams. In: Proc. ITS World congress, pp 1–8
3. Chen S, Sun ZP, Bridge B (1997) Automatic traffic monitoring by intelligent sound detection. In: Proc. IEEE conf. intelligent transportation systems (ITSC), pp 171–176
4. Chen S, Sun Z, Bridge B (2001) Traffic monitoring using digital sound field mapping. IEEE Trans Veh Technol 50(6):1582–1589
5. Forren JF, Jaarsma D (1997) Traffic monitoring by tire noise. In: Proc. IEEE conf. intelligent transportation systems (ITSC), pp 177–182
6. Ishida S, Liu S, Mimura K, Tagashira S, Fukuda A (2016) Design of acoustic vehicle count system using DTW. In: Proc. ITS World congress, pp 1–10. AP-TP0678
7. Ishida S, Mimura K, Liu S, Tagashira S, Fukuda A (2016) Design of simple vehicle counter using sidewalk microphones. In: Proc. ITS EU congress, pp 1–10. EU-TP0042
8. Knapp CH, Carter GC (1976) The generalized correlation method for estimation of time delay. IEEE Trans Acoust Speech Signal Process 24(4):320–327
9. Nurhadiyatna A, Hardjono B, Wibisono A, Jatmiko W, Mursanto P (2012) ITS information source: vehicle speed measurement using camera as sensor. In: Proc. int. conf. on advanced computer science and information systems (ICACSIS), pp 179–184
10. Taghvaeeyan S, Rajamani R (2014) Portable roadside sensors for vehicle counting, classification, and speed measurement. IEEE Trans Intell Transp Syst 15(1):73–83
11. Wu H, Siegel M, Khosla P (1998) Vehicle sound signature recognition by frequency vector principal component analysis. In: Proc. IEEE instrumentation and measurement technology conf. (IMTC), vol 1, pp 429–434

Algorithm/Architecture Codesign: From System on Chip to Internet of Things and Cloud

Gwo Giun (Chris) Lee, Chun-Fu Chen, and Tai-Ping Wang

1 Introduction

In the 1960s, Marshall McLuhan published the book entitled, "The Extensions of Man," focusing primarily on television, an electronic media as being the outward extension of human nervous system, which from contemporary interpretation marks the previous stage of big data.

In concurrent Industry 4.0 ecosystem, Internet of Things (IoT) facilitates extrasensory perception in reaching out even farther via sensors interconnected through signals with information exchange. Innovations in intelligent surveillance and monitoring technologies have not only made possible advancements toward smart cities, intelligent transportation systems (ITS) including autonomous cars, intelligent home (iHome), and intelligent biomedical and healthcare systems but also lead to the generation of even bigger data which will inevitably be witnessed. Further inward extension of human information perception could also be experienced when observing genomic, neurological, and other physiological phenomena when going deeper inward into the human body, again with tremendously big data such as from the human brain and especially the human genome.

Ubiquitous artificial intelligence (AI), brought forth by wearable, mobile, and other IoT devices, requires not only more complex algorithms but also automated analytics algorithm for versatile applications which start from science and engineering such as multimedia, communication, and biotechnology and will diversify toward other cross-disciplinary domains. Machine learning algorithms such as

G. G. (Chris) Lee (✉) · C.-F. Chen · T.-P. Wang
Department of Electrical Engineering, National Cheng Kung University, Tainan, Taiwan
e-mail: clee@mail.ncku.edu.tw

© Springer Nature Switzerland AG 2020
Y. Liu et al. (eds.), *Smart Sensors and Systems*,
https://doi.org/10.1007/978-3-030-42234-9_8

deep learning which have self-learning capabilities also demand excessively high complexity in processing these big heterogeneous data.

With mathematical fundamentals as foundations for the analysis of corresponding dataflow models from algorithms; intelligent, flexible, and efficient analytics architectures including both software and hardware for VLSI, GPU, multicore, high-performance computing, and reconfigurable computing systems; etc., this chapter innovates discussions on Smart system on chip design, in expediting the field of signal and information processing systems into futuristic new era of the Internet of Things and high-performance computing based on algorithm/architecture codesign.

2 Algorithm/Architecture Codesign: Analytic Architecture for Smart SoC

Niklaus Emil Wirth introduced the innovative idea that *programming = algorithm + data structure*. Inspired by this, we advance the concept to the next level by stating that *design = algorithm + architecture*. With concurrent exploration of algorithm and architecture entitled algorithm/architecture codesign (AAC), this methodology innovates a leading paradigm shift in advanced system design from system on a Chip to IoT and heterogeneous systems.

As high-performance computing becomes exceedingly demanding and IoT-generated data becomes increasingly bigger, flexible parallel/reconfigurable processing is crucial in the design of efficient and flexible signal processing systems with low power consumption. Hence the analysis of algorithms for potential computing in parallel, efficient data storage and data transfer is crucial. In analogous to the analysis of speech and image data in machine learning, this section characterizes the analysis of dataflow models representing algorithms, for analytics architecture, a cross-level-of-abstraction system design methodology for SoC on versatile platforms [1].

2.1 Architectural Platform

Current intelligent algorithms such as those for big data analytics and machine learning are becoming ever more complex. Rapid and continuous enhancements in semiconductor and information communication technologies (ICT) with innovations in especially advanced systems and architectural platforms capable of accommodating these intelligent algorithms targeting versatile applications including ubiquitous AI are therefore in high demand. These broad application-specific requirements such as for Smart SoC platforms necessitate trade-off among efficiency represented by performance per unit of silicon area (performance/silicon area), flexibility of usage due to changes or updates in algorithms, and low power consumption.

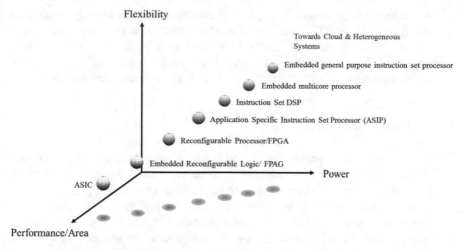

Fig. 1 Architectural platforms trading off performance/area, flexibility, and power

Conventional implementations of algorithms were usually placed at two architectural extremes of either pure hardware or pure software. Although application-specific integrated circuit (ASIC) implementation of algorithms provides the highest speed or best performance, this is however achieved via trade-off of platform flexibility. Pure software implementations on single-chip processors or CPUs are the most flexible, but require high power overhead and result in slower processing speed. Hence, several other classes of architectural platforms, such as instruction set digital signal processors (DSP) and application-specific instruction set processors (ASIP), have also been used as shown in Fig. 1.

It is thus crucial that system design methodologies, such as Smart SoC systems, emphasize optimal trade-off among efficiency, flexibility, and low power consumption. Consequently, embedded multicore processors or SoCs and reconfigurable architectures may frequently be favored. Furthermore, heterogeneous data generated from versatile IoT devices have further escalated system design toward cloud and heterogeneous systems in the post-Moore's law era.

2.2 Algorithm/Architecture Codesign: Abstraction at the System Level

As signal and information processing applications such as visual computing and communication become increasingly more complicated, the corresponding increase in hardware complexity in SoC design has also required reciprocity in software design especially for embedded multicore processors and reconfigurable platforms. In coping with large systems, design details for specific applications are abstracted into several levels of abstraction.

In traditional ASIC design flow, physical characteristics are typically abstracted as timing delay at the RTL level. For Smart SoC with yet even higher complexity, abstraction has been elevated further to system level with algorithmic intrinsic complexity metrics intelligently extracted from dataflow models, featuring both hardware and software characteristics for subsequent cross-level abstraction design.

2.2.1   Levels of Abstraction

The design space for a specific application is composed of all the feasible software and hardware implementation solutions or instances and is therefore spanned by corresponding design attributes characterizing all abstraction levels [2].

In a top–down manner, the design process in this method proceeds from algorithm development to software and/or hardware implementation. Abstracting unnecessary design details and separating the design flow into several hierarchies of abstraction levels as shown in Fig. 2 could efficiently enhance the design capability. For a specific application, the levels of abstraction include the algorithmic, architectural, register transfer, gate, and physical design levels. As shown in Fig. 2, more details are added as the design progresses to lower abstraction levels and hence with larger design space.

Figure 3 illustrates design details at every abstraction level of the design space. At the algorithmic level, functionalities are explored, and the characterizing time

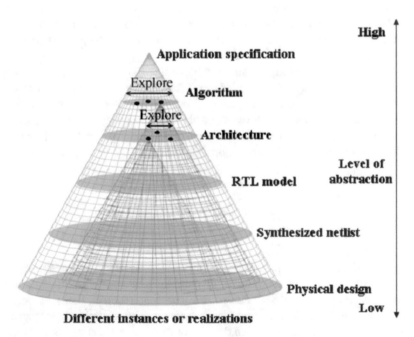

Fig. 2 Levels of abstraction

Levels	Symbols	Features	Time units
Algorithm	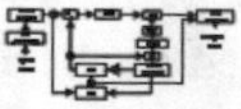	System functionality	Seconds
Architecture	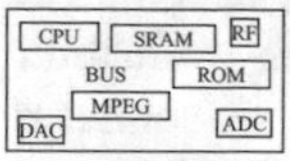 CPU SRAM RF BUS ROM MPEG DAC ADC	System architecture	Number of cycles
IP (Macro)	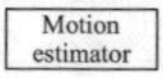 Motion estimator	IP functionality and micro-architecture	Number of cycles
Module	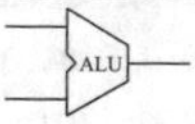 ALU	Arithmetic operation	Cycle
Gate	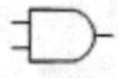	Logic operation	ns
Circuit	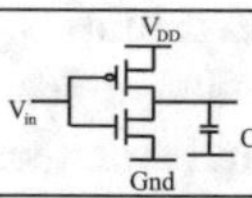 V_{DD} V_{in} C_L Gnd	Voltage, current	ps
Device	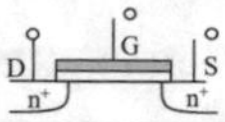 D G S n+ n+	Electron	ps

Fig. 3 Features at various levels of abstraction

unit used is in order of seconds. Real-time processing, for example, is a common constraint for visual applications, and the temporal domain precision is measured in terms of frames per second (FPS).

At the architectural level, exploration focuses on data transaction features including data transfer, storage, and computation. This information subsequently facilitates design for hardware/software partition, memory configuration, bus protocol, and modules comprising the system. The time unit is in the number of cycles.

At the silicon intellectual property (IP) or macro level, micro-architecture characteristics including the datapath and controller are considered, with the timing accuracy also counted in cycles. At the module level, features could, for instance, be various arithmetic units comprising the datapath. The gate level is characterized by logic operation for digital circuits. At the circuit level, voltage and current are notable. And finally, electrons are considered at the device level.

The discussions above reveal that higher levels of abstraction are characterized by coarser timing and physical scales and finer for lower levels. In traditional ASIC design flow, efforts were focused primarily at the register transfer level (RTL), where physical circuit behaviors with parasitical capacitance and inductance are abstracted within timing delay. In the currently proposed AAC design methodology, abstraction is further elevated to the system level where dataflow or transaction-level modeling bridges the cross-algorithm and architecture level design space.

2.2.2   Joint Exploration of Algorithms and Architecture

Traditional design methodologies are usually based on the execution of a series of sequential stages: the theoretical study of a fully specified algorithm, the mapping of the algorithm to a selected architecture, the evaluation of the performance, and the final implementation. However, these sequential design procedures are no longer adequate to cope with the increasing complexity demands of Smart SoC design challenges. Conventional sequential design flow yields independent design and development of the algorithm from the architecture. However, with ever-increasing complexity of both algorithm and system platforms in each successive generation, such unidirectional steps in traditional designs will inevitably lead to the scenario that designers may either develop highly efficient but highly complex algorithms that cannot be implemented or offer platforms that are impractical for real-world applications because the processing capabilities cannot be efficiently exploited by the newly developed algorithms. Hence, a seamless weaving of the previously autonomous algorithmic development and architecture development will unavoidably be observed.

As shown in Fig. 4, AAC facilitates the concurrent exploration of algorithm and architecture optimizations through the extraction of algorithmic intrinsic complexity measures from dataflow models. Serving as a bridge between algorithms containing behavioral information and architecture with design or implementation information, system-level features including the number of operations, degree of parallelism, data transfer rate, and data storage requirements are extracted as quantitative complexity measures to provide early understanding and characterization of the system architecture in cross-level designs.

As depicted in Fig. 2, the cost of design changes is high when designs have already progressed to the later stages at a lower level of abstraction and frequently affects the success of the overall project. Hence it is crucial that these algorithmic intrinsic complexity measures provide an early understanding of the architectural design and subsequent implementation requirements within the algorithm and architecture codesign space as shown in Fig. 5. This is in essence a systematic analytics architecture mechanism for the mapping of algorithms to platforms with the optimal balancing of efficiency, flexibility, and power consumption via architectural space exploration before software/hardware partitioning.

Fig. 4 Concept of algorithm/architecture co-exploration

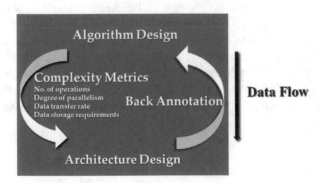

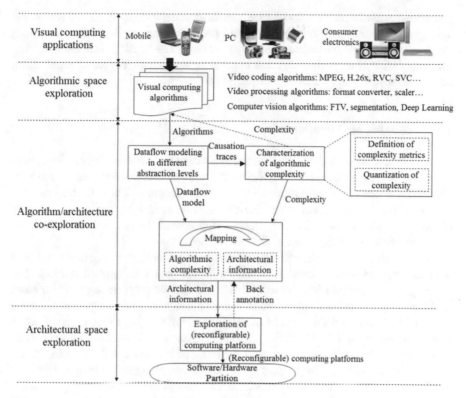

Fig. 5 Advanced visual system design methodology

Situations when the existing architectures or platforms are not able to accommodate the complexities that it is necessary to feedback or back-annotate the complexity information to the algorithmic level for algorithm modification are depicted in Figs. 4 and 5.

Hence AAC provides a cross-level methodology for smart system design by which abstraction of architecture features within complexity metrics has been further escalated to the system level! This is of course the same technique in traditional ASIC design flow with physical characteristics at physical layers being abstracted as timing parameters at the microarchitecture or RTL level.

2.3 Algorithmic Intrinsic Complexity Metrics and Assessment

Since algorithm/architecture co-exploration (AAC) is an iteration process of synthesis from algorithm to architecture and reconfiguration from architecture to algorithm, defining and extracting metrics of algorithms for processing AAC are crucial. The algorithmic intrinsic complexity metrics should not be biased toward software or hardware and should be platform independent so as to reveal the anticipated architectural features in AAC process. This section introduces four essential algorithmic intrinsic complexity metrics of Lee et al. [3] for characterizing the complexity of algorithms that are number of operations, degree of parallelism, data transfer rate, and data storage requirement.

2.3.1 Number of Operations

In terms of arithmetic and logic operators, evaluating the number of operations could represent the computational complexity of an algorithm because the number of operations is one of the most intuitive metrics of algorithms. The estimated number of operations of the algorithm can give designers an insight into either software or hardware development in early design stages, and it can effectively facilitate software/hardware partition and codesign.

To make the metric more accurate, we should consider the types of computational operations since various operations have different costs in implementation. The complexity of addition and subtraction is similar and simplest among the four basic arithmetic operators. Multiplication is more complex and can be performed by a series of additions and shifts based on Booth's algorithm [4]. Since division needs to be executed by shifts, subtractions, and comparisons, it is the most complicated.

Moreover, the precision of operand in terms of bit depth and type of operand (fixed point or floating point) also significantly influences the implementation cost and hence needs to be especially specified. In general, the gate count of processing elements increases as the precision grows higher. Besides, the hardware propagation delay is affected by the precision as well. If an algorithm is implemented on the processor-orientated platforms composed of general-purpose processors, single-instruction multiple-data machines, or application-specific processors, the precision of operand will directly determine the number of instructions needed to complete an operation. Consequently, the operand precision is also a very important parameter for measuring the number of operations.

2.3.2 Degree of Parallelism

The degree of parallelism is another metric characterizing the complexity of algorithms. Some partial operations within an algorithm are independent. These independent operations can be executed simultaneously and hence reveal the degree of parallelism. An algorithm whose degree of parallelism is higher has larger flexibility and scalability in architecture exploration. On the contrary, greater data dependence results in less parallelism, thereby giving a more complex algorithm. The degree of parallelism embedded within algorithms is one of the most essential complexity metrics capable of conveying architectural information for parallel and distributed systems at design stages as early as the algorithm development phase. This complexity metric is again transparent to either software or hardware. If an algorithmic function is implemented in hardware, this metric is capable of exhibiting the upper bound on the number of parallel PEs in the datapath. If the function is intended in software, the degree of parallelism can provide insight and hence reveal information pertaining to parallel instruction set architecture in the processor. Furthermore, it can also facilitate the design and configurations of multicore platforms.

One of the versatile parallelisms embedded within algorithms can be revealed as the independent operation sets that are independent of each other and hence can be executed in parallel without synchronization. However, the independent operation sets are composed of dependent operations that have to be sequentially performed. Hence, in a strict manner, the degree of parallelism embedded in an algorithm is equal to the number of fully independent operation sets. To efficiently explore and quantify such parallelism, Lee et al. [5] proposed to represent the algorithm by a high-level dataflow model and analyze the corresponding dataflow graph (DFG). The high-level dataflow model is capable of well depicting the interrelationships between computations and communications. The generated DFG can clearly reveal the data dependencies between the operations by vertexes and directed edges, where the vertexes denote the operations and the directed edges represent the sources and destinations of the data. Inspired by the principal component analysis in the information theory, Lee et al. [5] further employed the spectral graph theory [6] for systematically quantifying and analyzing the DFGs via eigendecomposition, so that the spectral graph theory can facilitate the analysis of data dependency and connectivity of the DFGs simplistically by means of linear algebra. Consequently, it is capable of quantifying the parallelism of the algorithm with robust, mathematical, and theoretical analysis applicable to a broad range of real-world scenarios.

Given a DFG G of an algorithm is composed of n vertexes that represent operations and m edges that denote data dependency and flow of data, in which the vertex set of G is $V(G) = \{v_1, v_2, \ldots, v_n\}$ and the edge set of G is $E(G) = \{e_1, e_2, \ldots, e_m\}$. The spectral graph theory can study the properties of G such as connectivity by the analysis of the spectrum or eigenvalues and eigenvectors of the Laplacian matrix $\mathbf{L}$ representing G, which is defined as [6, 7]:

$$\mathbf{L}(i, j) = \begin{cases} \text{degree}(v_i) & \text{if } i = j, \\ -1 & \text{if } v_i \text{ and } v_j \text{ are adjacent,} \\ 0 & \text{otherwise.} \end{cases} \quad (1)$$

where degree (v_i) is the number of edges connected to the ith vertex v_i. In the Laplacian matrix, the ith diagonal element shows the number of operations that are connected to the ith operation, and the off-diagonal element denotes whether two operations are connected. Hence, the Laplacian matrix can clearly express the DFG by a compact linear algebraic form.

Based on the following well-known properties of the spectral graph theory: (1) the smallest Laplacian eigenvalue of a connected graph equals 0 and the corresponding eigenvector $= [1, 1, \ldots, 1]^T$, (2) there exists exactly one eigenvalue $= 0$

for the Laplacian matrix of a connected graph, and (3) the number of connected components in the graph equals the number of eigenvalue $= 0$ of the Laplacian matrix; it is obvious that in a strict sense, the degree of the parallelism embedded within the algorithm is equal to the number of the eigenvalue $= 0$ of the Laplacian matrix of the DFG. Besides, based on the spectral graph theory, the independent operation sets can be identified according to the eigenvectors associated with the eigenvalues $= 0$.

This method can be easily extended to the analysis of versatile parallelisms at various data granularities, namely, multigrain parallelism. These multigrain parallelisms will eventually be used for the exploration of multicore platforms and reconfigurable architectures or instruction set architecture (ISA) with coarse and fine granularities, respectively. If the parallelism is homogeneous at fine data granularity, the single-instruction multiple-data (SIMD) architecture is preferable, since the instructions are identical. On the contrary, the very long instruction word (VLIW) architecture is favored for dealing with the heterogeneous parallelism composed of different types of operations. As the granularity goes coarser, the types of parallelism can help design the homogeneous or heterogeneous multicore platforms accordingly. In summary, this method can efficiently and exhaustively explore the possible parallelism embedded in algorithms with various granularities. The multigrain parallelism extracted can then facilitate the design space exploration for the advanced AAC.

By directly setting eigenvalues of $\mathbf{L} = 0$, it is easy to prove that the degree of parallelism is equal to the dimension of the null space of $\mathbf{L}$ and the eigenvectors are the basis spanning the null space. In general, the number of operations needed to derive the null space of a Laplacian matrix is proportional to the number of edges. Hence, this method provides an efficient approach to quantify the degree of parallelism and the independent operation sets. This method is applicable to large-scale problems by avoiding the computation-intensive procedures of solving the traditional eigendecomposition problem. In addition, since the Laplacian matrix is sparse and symmetrical, it can be efficiently implemented and processed by linking list or compressed row storage (CRS) format.

2.3.3 Storage Configuration

A system is said to be memoryless if its output depends on only the input signals at the same time. However, in visual computing applications such as video coding and processing, some intermediate data have to be stored in memory depending on the dataflow of algorithms in higher abstraction levels. Consequently, in order to perform the appropriate algorithmic processing, data storage must be properly configured based on the dataflow scheduling of the intermediate data. Hence, the algorithmic storage configuration is another essential intrinsic complexity metric in AAC design methodology, which is transparent to either software or hardware designs. For software applications, the algorithmic storage configuration helps in designing the memory modules such as cache or scratch pad and the corresponding data arrangement schemes for the embedded CPU. In hardware design, the immediate data can be stored in local memory to satisfy the algorithmic scheduling based on this complexity metric.

To provide a better visual quality in visual computing, for example, more context information should be stored to exploit, and hence the storage size requirement is intended to be increased. Usually, the picture data is stored in the external storage due to the large amount of data. Therefore, the data transfer rate balance between internal and external storage is crucial. There are two extreme cases of this consideration: (1) All the needed data is stored in the internal storage that requires the minimum external data transfer rate and (2) all the required data is stored in the external storage that requires the maximum external data transfer rate since the needed data would be fetched when the algorithm demands. An intuitive manner is to allocate partial picture data in the internal storage and remaining data in the external storage. However, these two factors are usually inversely proportional. In the following, a systematic manner to explore the balance between internal data storage and external data transfer rate through different executing orders is demonstrated. Hence, a feasible solution can be found during the design space exploration for the target application of multidimensional video signal processing.

An executing order in dataflow affects internal storage size and external data transfer rate, and the executing order is always restricted to the data causality of the algorithm. Figure 6 shows a dataflow dependency graph of a typical image/video processing algorithm exploiting the contextual information in the spatial domain. To a causal system, only upper and left contextual information can be referenced.

Three different executing orders are illustrated in Fig. 7, including (a) the raster scan order, (b) diagonal scan order with two rows, and (c) diagonal scan order with three rows, and the number labeled on the vertices denotes the executing order of nodes. There are some assumptions applied for discussing the effect of executing

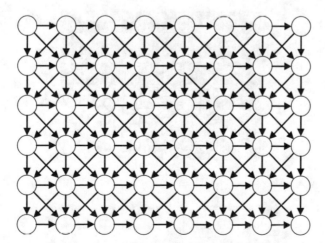

Fig. 6 Dataflow dependency graph of a typical image/video processing algorithm

order on the internal data storage and the external data transfer rate. The contextual information at left side is stored in the internal storage, and the data at upper line should be fetched from external storage. Thus, the internal storage size is counted according to the data size of left reference, and the average external data transfer rate is measured based on the amount of upper data reference which should be fetched within one time unit.

By analyzing the dataflow illustrated in Fig. 7a, the required storage size is one data unit, and the external data transfer rate is three data units. The dataflow depicted in Fig. 7b needed to store three data units and transfer three data units during the processing of every two data units. The last one dataflow illustrated in Fig. 7c stored five data units, and three data units should be transferred when processing every three data units.

In summary, the first dataflow requires the smallest data storage requirement, but the average data transfer rate is the largest among these three dataflow models due to the fact that the required data would be fetched from external data storage

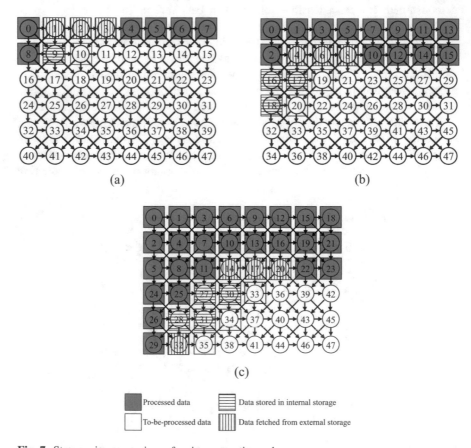

(a)   (b)

(c)

Fig. 7 Storage size comparison of various executing orders

Fig. 8 Normalized average external data transfer rates versus internal storage sizes for various executing order

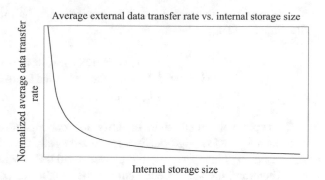

Average external data transfer rate vs. internal storage size

once requisition. On the other hand, the third dataflow possesses the largest internal storage size since more contextual information should be kept to process the data unit at distinct rows; however, the required average data transfer rate is the smallest one because most of the data have been stored in the internal storage already.

The trade-off between internal storage size and average data transfer rate is made in accordance with the distinct executing orders. Figure 8 shows the analyzed result of the diagonal scan from 1 row to 32 rows. The normalized average data transfer rate is inverse proportional to the internal data storage size. Figure 8 shows that the reduction ratio of average data transfer rate could be achieved by adding some overhead on the internal storage size. The curve in Fig. 8 can facilitate the design space exploration in terms of the internal data storage and external data transfer rate based on AAC.

2.3.4 Data Transfer Rate

In addition to the number of operations, degree of parallelism, and storage configuration, the amount of data transfer is also an intrinsic complexity metric as executing an algorithm. Algorithms can be represented by natural languages, mathematical expressions, flowcharts, pseudocodes, high-level programming languages, and so on. In signal processing applications, mathematical expression is one of the most abstract, definite, and compact methods to represent an algorithm. The corresponding signal flow graphs and dataflow models [2, 8] can be then obtained based on the mathematical representation [9]. The dataflow model is capable of depicting the interrelationships between computations and communications.

To systematically extract the information embedded in the graph, matrix representation is commonly used to represent a DFG. For instance, the adjacent matrix introduces the connections among vertices, and the Laplacian matrix also displays the connectivity embedded in the graph. These matrix representations are usually in behalf of undirected graphs; however, in the study of data transfer of visual signal processing, data causality is also significant information that should be retained in matrix representation. Hence, a dependency matrix conveying data causality of a directed or undirected graph is required, and its mathematical expression is illustrated as (2):

$$\mathbf{M}(i, j) = \begin{cases} -1, & \text{if vertex } v_j \text{ is the tail of edge } e_i \\ 1, & \text{if vertex } v_j \text{ is the head of edge } e_i \\ 0, & \text{otherwise} \end{cases} \tag{2}$$

To explore the method to quantify the corresponding data transfer rate via dependency matrix, edge cut is applied since edge cut is a cut that results in a connected DFG into several disconnected sub-DFGs by removing the edges in this cut. Therefore, the size of edge cut (or the number of edges in this cut) could be used to estimate the amount of data that would be transferred among sub-DFGs due to the fact that data should be sent or received (via edges) by tasks (vertices). On the other hand, the behavior of edge cut in DFG is equivalent to applying an indicator vector $\mathbf{x}$ that separates vertices in DFG into two sides for the dependency matrix, $\mathbf{M}$. Furthermore, by computing $\mathbf{Mx}$, the characteristics of edges in DFG would be revealed. According to $\mathbf{Mx}$, the amount of data transfer was equal to half of the summation of all absolute values in $\mathbf{Mx}$. Therefore, $\mathbf{Mx}$ clearly presents the number of edges crossed by this edge cut, and hence corresponding data transfer rate could be systematically quantified due to the fact that data transactions occurred on the edges in DFG.

For example, a simple DFG of an average filter is shown in Fig. 9. The indicator vector $\mathbf{x}$ of corresponding edge cut is $[1, -1, -1, 1]^T$; this edge cut separates v_1 and v_4 into one group and v_2 and v_3 into the other group. (The vertices at the side with more input data would be set as 1.) In this example, $\mathbf{Mx}$ is $[2, 0, -2]^T$, and there are three types of edges that are introduced by $\mathbf{Mx}$, including in-edge-cut (positive value in $\mathbf{Mx}$, e_1), out-edge-cut (negative value in $\mathbf{Mx}$, e_3), and non-edge-cut (zero value in $\mathbf{Mx}$, e_2). The corresponding dependency matrix ($\mathbf{M}$), indicator vector ($\mathbf{x}$), characteristics of edges ($\mathbf{Mx}$), and amount of data transfer are depicted in (3). Consequently, the amount of data transfer of this edge cut is 2.

Fig. 9 A simple DFG and an edge cut separate vertices into two sides

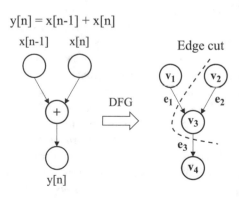

$y[n] = x[n-1] + x[n]$

$x[n-1]$ $x[n]$

Edge cut

DFG

$$\mathbf{M} = \begin{matrix} & v_1 \ v_2 \ v_3 \ v_4 \\ e_1 \\ e_2 \\ e_3 \end{matrix} \begin{bmatrix} 1 \ 0 \ -1 \ 0 \\ 0 \ 1 \ -1 \ 0 \\ 0 \ 0 \ 1 \ -1 \end{bmatrix}, \mathbf{x} = \begin{bmatrix} 1 \\ -1 \\ -1 \\ 1 \end{bmatrix} \Rightarrow \mathbf{Mx} = \begin{bmatrix} 2 \\ 0 \\ -2 \end{bmatrix} \tag{3}$$

$$\text{Amount of data transfer} = \tfrac{1}{2} \sum_{i=1}^{N} |\mathbf{Mx}(i)| = 2$$

3 Toward Big Data Analytics in SMARTECH/AI with Analytic Architecture for Edge/Fog/Cloud Computing

As discussed in previous sections, AAC can not only be used in hardware design but also be suitable for software design and edge/fog/cloud computing deployment. Deploying cloud computing for AI development needs to consider distributing compute, data bandwidth, and storage. On edge computing, we should take data reusability and computation saving into consideration. AAC could help us find out optimized solutions for edge/cloud computing. This section states what parallel/reconfigurable computing in AAC is and gives an edge/fog computing by AAC method.

3.1 Intelligent Parallel/Reconfigurable Computing

In signal processing, Fourier transformation is adopted to analyze and synthesize signals. AAC also provides a similar framework called parallel and reconfigurable computing in analytic architecture. The eigen analysis of dataflow graphs and graph component synthesis in AAC are for parallel and reconfigurable computing.

AAC presents a spectral graph theory technique which systematically lays out the full spectrum of potential parallel processing components eigendecomposed into all possible data granularities. This makes possible the study of both quantitative and qualitative potentials for homogeneous or heterogeneous parallelization at different granularities as opposed to systolic array for homogeneous designs at only one single fixed granularity. In addition, the capabilities of AAC also include facilitating systematic analyses of dataflow models for flexible and efficient data transfer and storage.

Reconfigurable architectures including multicore and GPU platforms provide a balance between flexibility, performance, and power consumption. Starting from algorithms, the data granularity could be reduced so as to extract common functionalities among different algorithms. To reduce the granularity from the architectural side, the eigendecomposition of dataflow models described above could also be used to decompose connected graphs to disconnect components with different granularities. These commonalities would then require one design of either software or hardware which could be shared. These lower granularity commonalities also provide quantitative guidance in reconfiguring architectural resources such as in multicores or GPUs through graph component synthesis.

3.2 Edge/Fog Computing

Deep learning drew significant attention in the machine learning community around one decade ago [10], especially in computer vision area. Deep convolutional neural network (CNN) exploits hierarchical visual characteristics through very deep neural network architecture to achieve impressive performance in several visual applications, such as facial recognition [11], image classification [12], object detection [13], etc. In many applications, only a fixed trained model is required to inference or predict the most probable class of test targets. On the other hand, the time-consuming part, training deep CNN model, can be completed offline through high-performance computing machines. For instance, we can train a deep CNN model via a high-performance CPU/GPU cluster and then deploy the trained model onto edge/mobile devices for performing inference or build cloud applications to achieve machine intelligence.

To achieve device intelligence, there are two typical ways with respect to where the computation is completed: (1) cloud-based solution, which uploads data to the cloud and then receives results, and (2) edge computation, which processes through edge computation engine. There are three major weaknesses for the cloud-based solution, including latency, privacy, and power consumption. The cloud-based solution needs to send/receive data to/from the cloud which depends on the network environment, and the data might be leaked during transferring. Furthermore, cloud-based inference requires more than two times power as compared to processing on edge/mobile devices [14]. In contrast, edge computing can provide promising on-device inference without the above issues.

For deep CNN inference engine on edge computing, two features are included to resolve heavy computation load in a convolution layer: (1) high data reusability for low power and highly effective bandwidth for parallel computing and (2) convolution kernel redundancy removal to save computations and storage size. Based on algorithm/architecture co-exploration [1], we explore dataflows with different sizes of computing block (CB), where CB denotes 3-D data volume of produced outputs, to exploit the degree of data reusability to maximize the equivalent data transfer rate. Therefore, with a highly effective data transfer rate, every computing thread could get data on time to achieve high throughput. Furthermore, since deep CNN is usually over-parametrized, we reduce computation complexity and model size via exploiting redundant convolution kernels.

3.2.1 Data Reusability Exploration

On edge devices, we need to carefully utilize resources to maximize computing power or throughput, e.g., we would like to reuse data as much as possible to reduce power consumption in data loading. In our proposed data locality-awareness computing system, we exploit the data locality to maximize the data reusability and parallel computing; therefore, we leverage the memory bandwidth to parallelize convolution, which requires massive data loading operations.

In convolution layers, every kernel loads an entire input map to produce one output map which results in massive data transfer. In a naive approach and without considering data reuse, to produce a point in output map, it requires loading coefficients of convolution kernels and corresponding input data; therefore, it results in $2 \times C \times H \times W$ data transfer, where C, H, and W are channel, height, and width of one convolution kernel. To generate all output maps ($N \times Y' \times X'$), the amount of data transfer needed is $(2 \times C \times H \times W) \times (N \times Y' \times X')$, where N, Y', X' are channel, height, and width of the output tensor. On the other extreme, we can only load input map and all convolution kernels once and then generate whole output maps; in this case, only $(C \times X \times Y) + (N \times C \times H \times W) + (N \times Y' \times X')$ data transfer is required which is significantly lower than the naive approach, but it leads to a larger storage size to buffer the entire input map and convolution kernels, where C, Y, and X are channel, height, and width of the input tensor. Similar results can be conducted in the pooling layer and the fully connected layer.

Therefore, we explore different dataflows at various CBs to reuse data among all computing threads to resolve the very high bandwidth requirement for parallel computing based on previous work [15]. The dataflow of a convolution layer is regular, and we can analyze data reusability via mathematical expressions directly. With different sizes of CB, we explore the degree of data reusability to find out suitable CB for our target systems.

Data reuse happens when moving from the current CB to the next CB. Since the current CB is usually right next to the next CB, partial data are shared among two consecutive CBs. We do not need to load all data required by the next CB if we reuse common data used by the current CB and the next CB. To simplify the analysis, we assume that the first CB starts from the top-left of an image and goes with raster scan order and different processing flows can be analyzed in similar ways. The degree of data reusability is defined as the number of reusable data divided by number of generated output data, and it can be formulated as $((W - 1) \times (CB_y + H - 1) \times C + W \times H \times C)/(CB_x \times CB_y \times CB_z)$, where numerator is to-be-updated data when moving from the current CB to the next CB and CB_i is the CB size along i-axis (i can be x, y, or z, which denotes the three axes of a tensor, respectively.). Figure 10 displays data reusability at various CBs and kernel sizes, and the value on the y-axis is normalized to the case of $CB_{4 \times 4 \times 32}$; this analysis also shows that the ratio of data reusability keeps consistent trend no matter what the kernel sizes are. Therefore, it is a generic approach to reduce bandwidth via reusing data as much as possible; however, the analyzed results show that we achieved high data reusability with a small buffer, e.g., when comparing

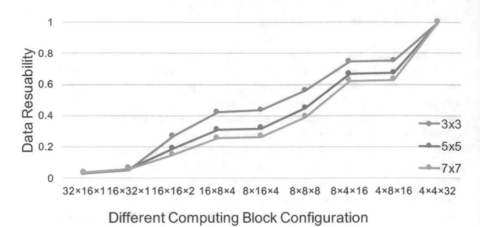

Fig. 10 Degree of data reusability at various computing block sizes and kernel sizes. The highest reusability is normalized to 1 and the total computing thread is 512

$CB_{32 \times 16 \times 1}$ and $CB_{4 \times 4 \times 32}$ at 3×3 kernel size, $CB_{4 \times 4 \times 32}$ can reuse 31.5 times the data with only 2.3 times the buffer size.

On the other hand, temporary buffers stored intermediate features (output of every layer) in deep CNN that can be reused easily via a ping-pong buffer strategy since lifetime of all intermediate features immediately finish after they are consumed; therefore, we can just use two buffers instead of multiple buffers to save storage size.

3.2.2 Computation Saving

Deep CNN usually uses over-parametrized models to extract features with thousands of convolution kernels; therefore, lots of kernels might contribute little or none at all. This situation can be observed while the CNN is very deep such as VGG16 or VGG19, which are developed by Visual Geometry Group (VGG). The weight sparsity at deeper convolution layers is usually high since deeper layers are used to extract specific features. Therefore, we can exploit kernel redundancy to reduce computation complexity and storage requirement for deep CNN.

The weight distributions of convolution layers in VGG16 are like single-side Laplace distribution with different scales, and the scales of the first three layers are larger than the last three layers. Lots of weights at the last three convolution layers are close to zero; in other words, weights could be sparse. Hence, lots of kernels do not extract features, so we can discard them with slight performance degradation to reduce the computation complexity. We prune redundant kernels via exploiting the sparsity of each kernel. Subsequently, after pruning, we need to fine-tune the model to transfer the features learned by pruned kernels to remained kernels. We define the sparsity of the nth convolution kernel as (4), where th_1 is the threshold considering the weights are ignorable. Then, if a kernel has several coefficients less than th_1, $Sparsity(n)$ is close to 1 and it means this kernel is sparse. Afterward, we evaluate sparsity via another threshold, th_2, to determine which kernel should be discarded, since if most coefficients in one kernel are close to zero, it also denotes this kernel is unused.

$$Sparsity(n) = \frac{\sum_c^C \sum_j^H \sum_i^W \sigma(W(n,c,j,i))}{C \times W \times H}$$
$$\sigma(x) = \begin{cases} 1, \text{ if } |x| < th_1 \\ 0, \text{ otherwise} \end{cases} \tag{4}$$

The advantage of this pruning approach is the entire computation flow does not change, but the number of kernels at a convolution layer is reduced. Therefore, we do not need to redesign our computation engine of a convolution layer, and the speedup is also linearly proportional to the reduced operations. Furthermore, by reducing kernel numbers, we can also gain the benefits of lower storage requirements for trained kernels. With fewer convolution kernels, the feature dimension also decreases, and it results in the weights between the last convolution layer and the first fully connected layer to become smaller.

References

1. Lee GG, Chen YK, Mattavelli M, Jang ES (2009) Algorithm/architecture co-exploration of visual computing: overview and future perspectives. In: IEEE transactions on circuits and systems for video technology, vol. 19, pp 1576–1587
2. Edwards S, Lavagno L, Lee EA, Sangiovanni-Vincentelli A (1997) Design of embedded systems: formal models, validation and synthesis. Proc IEEE 85(3):366–390
3. Lee GG, Lin H-Y, Kung S-Y (2012) Algorithm/architecture co-exploration. In: Multimedia image and video processing, vol 573–608, 2nd edn. CRC Press, Boca Raton
4. Booth AD (1951) Signed binary multiplication technique. Q J Mech Appl Math 4(2):236–240
5. Lin HY, Lee GG (2010) Quantifying intrinsic parallelism via eigen-decomposition of dataflow graphs for algorithm/architecture co-exploration. In: Proceedings of IEEE SiPS 2010
6. Chung FRK (1997) Spectral graph theory. In: Regional conferences series in mathematics, no 92
7. Fiedler M (1973) Algebraic connectivity of graphs. Czechoslovakia Math J 23(2):298–305
8. Ragan-Kelley J, Barnes C, Adams A, Paris S, Durand F, Amarasinghe S (2013) Halide, a language and compiler for optimizing parallelism, locality, and recomputation in image processing pipelines. In: Proceedings of the 34th ACM SIGPLAN conference on programming language design and implementation, Seattle, Washington
9. Oppenheim AV, Schafer RW (1989) Discrete-time signal processing. Prentice-Hall, Englewood Cliffs
10. Hinton G, Osindero S, Teh Y (2006) A fast learning algorithm for deep belief nets. Neural Comput 18(7):1527–1554
11. Taigman Y, Yang M, Ranzato M, Wolf L (2014) DeepFace: closing the gap to human-level performance in face verification. In: 2014 IEEE conference on computer vision and pattern recognition, pp 1701–1708
12. Szegedy C, Ioffe S, Vanhoucke V (2016) Inception-v4, inception-ResNet and the impact of residual connections on learning. CoRR vol. abs/1602.07261

13. Ren S, He K, Girshick R, Sun J (2015) Faster R-CNN: towards real-time object detection with region proposal networks. Adv Neural Inf Proces Syst 28:91–99
14. Lane ND, Bhattacharya S, Georgiev P, Forlivesi C, Jiao L, Qendro L, Kawsar F (2016) DeepX: a software accelerator for low-power deep learning inference on mobile devices. In: 2016 15th ACM/IEEE international conference on information processing in sensor networks (IPSN). IEEE, pp 1–12
15. Chen CF, Lee GGC, Yu ZH, Huang CH (2014) Mapping visual signal processing onto multi-core platform via algorithm/architecture coexploration. In: 2014 IEEE workshop on signal processing systems (SiPS), pp 1–6

Engineering Elastomer Dielectric for Low-Cost and Reliable Wearable Health and Motion Monitoring

Sujie Chen, Siying Li, Yukun Huang, Jiaqing Zhao, Wei Tang, and Xiaojun Guo

1 Introduction

Thin and flexible sensors directly attached on the skin surface or cloths for real-time monitoring of diverse human physiological signals and body motions have attracted wide attention for applications in wearable healthcare and patient rehabilitation [1, 2]. For example, as illustrated in Fig. 1, with the strain and pressure sensors, various physiological signals related to health conditions and body motions can be monitored [3, 4]. To be able to detect these different physical signals in real time, the sensors need to achieve the required limit of detection, detection range, and sensitivity with fast enough response and recovery. For long-term wearable applications, biocompatibility with human skin, low power consumption, and durability of sensing performance are also key prerequisites. Another important feature, which would determine wide adoption of the sensors, is being imperceptible when they are worn on the human body. For that, the sensors need to be ultrathin, light, and of low Young's modulus. Moreover, in some application cases (e.g., sensors on exposed parts of the skin), transparency or semitransparency is preferred for the sensors to be invisible [5]. Currently, polydimethylsiloxane (PDMS) elastomer is the popularly used base material for constructing various kinds of pressure and strain sensors due to its excellent flexible and elastic properties, biomedical compliance with human tissues, and commercial availability [6–9]. With external pressure or strain being applied, the induced mechanical structure deformation of PDMS (e.g., length or film thickness) results in changes of the resistance or capacitance of the fabricated devices to measure the applied stimuli [10, 11]. Therefore, the sensing performance is determined by the mechanical properties of the elastomer film. Unsatisfactorily,

S. Chen · S. Li · Y. Huang · J. Zhao · W. Tang · X. Guo (✉)
Department of Electronic Engineering, Shanghai Jiao Tong University, Shanghai, China
e-mail: x.guo@sjtu.edu.cn

© Springer Nature Switzerland AG 2020
Y. Liu et al. (eds.), *Smart Sensors and Systems*,
https://doi.org/10.1007/978-3-030-42234-9_9

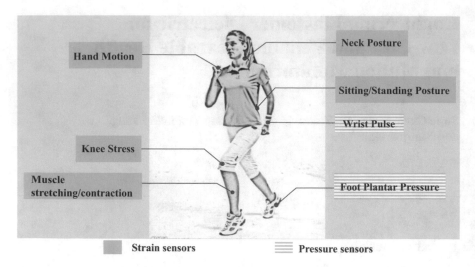

Fig. 1 Illustration of using the strain and pressure sensors to monitor various physiological signals related to health conditions and body motions

the bulk PDMS film, having a modulus of ~1 MPa, cannot generate large enough structure deformation to be converted to detectable signal changes under weak pressure or strain forces [12]. Structuring the PDMS films to reduce the modulus for improved sensitivity has become a research hotspot [6, 13–18]. On the other hand, to obtain light and imperceptible sensors, processes and designs of making devices on ultrathin PDMS film are needed [19].

This chapter will firstly review the device structures based on PDMS elastomer for both pressure and strain sensors. Then, the developed low-cost processing techniques for microstructuring PDMS films and making ultrathin PDMS films will be introduced. Pressure and strain sensors are fabricated with these films and integrated in wearable systems for monitoring diverse human physiological signals and body motions, including wrist pulse monitoring, static and dynamic foot pressure monitoring, and neck posture monitoring.

2 Device Structures

The pressure and strain sensors can be both constructed based on either capacitor or resistor structures using PDMS elastomer. With the capacitor structure, the PDMS layer is sandwiched between two electrodes (Fig. 2a). For pressure sensing, an external pressure is applied onto the surface of the top electrode, and the induced deformation of the dielectric layer (e.g., reduction in the dielectric layer thickness) causes changes of the capacitance to be measured. With the similar capacitor structure, if a strain force is applied along the transverse direction, the resulted

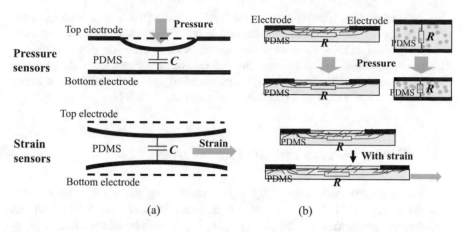

Fig. 2 Illustration of structures for constructing both pressure and strain sensors based on PDMS films: (**a**) capacitor structure and (**b**) resistor structure

change of the dielectric thickness can also cause the capacitance change to be detected for strain sensing [20].

For the different resistor structures as shown in Fig. 2b, the active part between two electrodes uses PDMS elastomer mixed with different forms of conducting materials, which could be powders (e.g., carbon black), or 0D/1D/2D nanostructure materials (nanoparticles, carbon nanotubes, metal nanowires, graphene, etc.). These conductive materials form conductive paths in the insulating host polymer. For pressure sensing, the deformation of the PDMS layer enhances the contacts or shortens the distance among the conductive materials and in turn causes the measured resistance change. For strain sensing, the deformation of the PDMS film (e.g., change in the length) upon the applied strain force affects the connections of the conductive materials, resulting in the measured resistance change as well.

For these device structures, the mechanical properties of the PDMS layer play an important role of determining the sensing performance. To improve the sensitivity, it is needed to decrease the modulus of the PDMS layer. In the following two sections, two approaches including microstructuring PDMS film and making ultrathin films will be introduced for this purpose.

3 Microstructured PDMS for Pressure Sensors

For the capacitor structure sensors in Fig. 2a, the bulky PDMS elastomer film is not able to produce enough deformation upon very small compression to obtain detectable capacitance change [10]. As a result, the fabricated sensors are not applicable for reliable sensing of weak physiological signals such as wrist pulse. Moreover, the viscoelastic behavior of unstructured PDMS thin film might cause slow relaxation time after removal of the applied force [10], which is an issue for real-time applications. To address these problems, different approaches for creating microstructured PDMS films were developed, with formed air voids inside the film to reduce elastic resistance and also induce additional change of the effective dielectric constant of the dielectric layer upon applied pressure for higher sensitivity since air has a lower dielectric constant ($\varepsilon = 1.0$) than PDMS ($\varepsilon \sim 3.0$) [21, 22].

3.1 Microstructuring with 3D Printed Mold

The most straightforward approach for microstructuring the PDMS film is using a prefabricated mold. Silicon wafer molds were fabricated to obtain different well-defined geometries for the purpose [13–17]. However, this method requires expensive and complicated photolithography and chemical etching processes and is thus not applicable for low-cost large area manufacturing.

Three-dimensional (3D) digital printing, which has recently been popularly used for rapid prototyping, is proposed as an alternative choice to replace photolithography for low-cost and convenient mold fabrication [21]. As illustrated in Fig. 3a, a mold of 3 cm × 3 cm size with periodical microgrooves on the surface was fabricated by a 3D printer (UP Plus 2 from Tiertime) using acrylonitrile butadiene styrene (ABS). The periodical microgrooves are of 100 μm in depth and 150 μm in width with the period width of 400 μm. The period width is designed as large as possible for maximized air-void ratio in the replicated PDMS film based on the mold. A prepared 10:1 mixture of PDMS elastomer (Sylgard 184, Dow Corning)

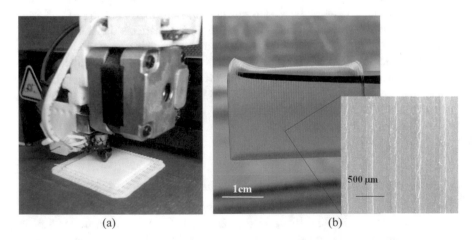

(a)   (b)

Fig. 3 (a) Photo image of the 3D printer (UP Plus 2 from Tiertime) used for fabricating the mold based on acrylonitrile butadiene styrene (ABS). (b) Photo image of the fabricated freestanding PDMS film (3 cm × 3 cm) with the scanning electron microscope (SEM) showing the surface structure. (Edited from [21])

to cross-linker was cast onto the fabricated mold and degassed in the vacuum chamber for 10 min, followed by a curing process at 65 °C for 60 min. The freestanding PDMS film with microstructured surface was peeled off from the mold and strengthened by another curing process at 100 °C for 60 min. The photo image of the obtained microstructured PDMS film is shown in Fig. 3b with the surface image taken by scanning electron microscope (SEM). The formed periodical micro-strips on the surface have a width of about 129 μm and a period width of about 550 μm, resulting in a very high air-void ratio of about 94% in the layer. When an external pressure is applied onto the fabricated sensor, the high air-void ratio will help to induce large deformation and change of the effective dielectric constant and in turn large capacitance change for high sensitivity in the low-pressure regime. The high air-void ratio in the PDMS layer also helps to reduce the relaxation time after removal of the pressure load, and fast response can thus be achieved. In the structure, the micro-strip width can be further reduced by using a higher-resolution 3D printer to fabricate the mold, so that the period width can be decreased while maintaining the similar air-void ratio.

To test its application for pressure sensors, the microstructured PDMS film was laminated onto an indium tin oxide (ITO)-coated poly(ethylene terephthalate) (PET) film, which was then cut into smaller pieces for use. The sensor devices were constructed by laminating two pieces of PDMS/ITO/PET films face-to-face with the micro-strip structures on two PDMS films perpendicularly crossing as shown in Fig. 4a. Electrical contacts were made by attaching conductive copper tapes (3 M 1811) onto the ITO electrodes. The photo image of the finished flexible sensor device is shown in Fig. 4b. The measured relative capacitance change ($\Delta C/C_0$) as a function of the applied pressure (P) for the sensor device is shown in Fig. 4c. C_0 is the capacitance when no pressure is applied on the capacitance (5.15 pF in this work), and ΔC is the measured capacitance change upon external applied pressure. The pressure sensitivity S is defined as $S = \delta(\Delta C/C_0)/\delta P$. It can be seen from Fig. 4c that the sensor exhibits a high sensitivity of 1.62 kPa^{-1} in the low-pressure regime (<0.2 kPa). For monitoring human physiological signals, the devices are operated in the low-pressure regime. As compared in the inset of Fig. 4c, the achieved sensitivity in the low-pressure regime with the device is higher than those of previous work using silicon molds made by micro-fabrication processes. The sensor also presents fast response and recovery in the millisecond range upon loading and unloading external pressure of different levels, as shown in Fig. 4d, e, which indicates its capability for real-time pressure monitoring applications. The operational stability of the sensor was characterized by applying more than 1000 cycles of repeated loading/unloading a pressure of 1 kPa for 2500 s, as shown in Fig. 4f. The device presented excellent stability with very small $\Delta C/C_0$ variation less than 2% during the test.

Finally, as a practical application example for monitoring human physiological signals, the sensor was connected to the self-designed data acquisition (DAQ) circuit board, which was powered by a 3.3 V voltage supply using a Li-ion battery and able to acquire the measured capacitance value from the connected sensor for further process, and sent the data to a mobile phone wirelessly through Bluetooth. As shown

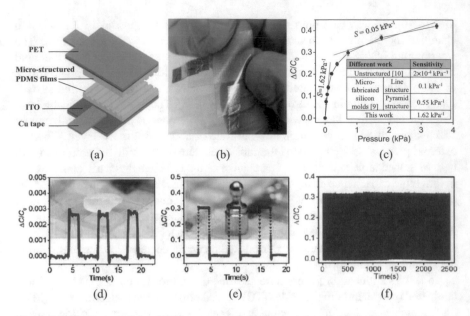

Fig. 4 (**a**) Illustration of the device structure of the fabricated capacitive pressure sensors and (**b**) photo image of the finished device in a size of 1.5 cm × 1.5 cm. (**c**) The measured relative capacitance change ($\Delta C/C_0$) as a function of the applied pressure for the sensor device. Inset: comparison of the measured sensitivity in the low-pressure regime (<0.2 kPa) with previously reported flexible capacitive sensor with unstructured PDMS layer and microstructured PDMS layers using micro-fabricated silicon molds. Dynamic response of the sensor upon placement and removal of (**d**) a rice grain (15 mg, corresponding to the pressure of 3 Pa) and (**e**) a weight (5 g, corresponding to the pressure of 1 kPa). (**f**) Characterization results of the operational stability of the sensor with continuous measurement of relative capacitance change ($\Delta C/C_0$) for 2500 s by applying more than 1000 cycles of repeated loading/unloading a pressure of 1 kPa. (Edited from [21])

in Fig. 5a, the whole system was placed on the radial artery of the wrist and fixed by transparent adhesive tape. Continuous real-time monitoring of wrist pulse would thus provide a convenient and noninvasive way for diagnosing medical diagnosis [23]. The upper part of Fig. 5b shows the real-time measured relative capacitance change $\Delta C/C_0$ for the volunteer. The full wrist pulse contour can be clearly detected with the wearable sensor system, showing regular and repeatable pulse shape with a pulse frequency of about 79 beats/min. The lower part of Fig. 5b presents the details of the pulse wave in one period, which contains the percussion wave (P-wave), the tidal wave (T-wave), and the diastolic wave (D-wave). These three different parts are related to the systolic and diastolic blood pressure, the ventricular pressure, and the heart rate, respectively, and can provide important clinical information [23]. The results demonstrate the capability of the fabricated pressure sensor to be integrated in a wearable system for monitoring the weak human physiological signals in real time.

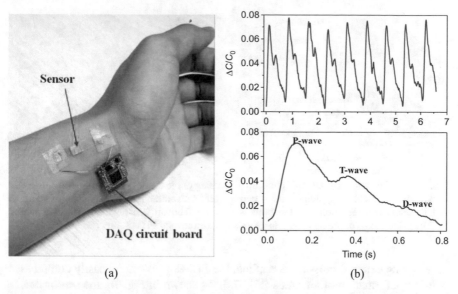

(a)   (b)

Fig. 5 (**a**) The whole sensor system, composed of the pressure sensor and the self-designed data acquisition (DAQ) circuit board, is placed on the radial artery of the wrist and fixed by transparent adhesive tape for real-time monitoring of wrist pulse. (**b**) The real-time measured wrist pulse waveform with relative capacitance change $\Delta C/C_0$ (upper), and the details of the pulse wave in one period (bottom), which contains the percussion wave (P-wave), the tidal wave (T-wave), and the diastolic wave (D-wave). (Edited from [21])

3.2 Porous PDMS Films

A more facile approach is developed for one-step processing of large area microstructured elastomer film with high-density micro-features of air voids and can be seamlessly integrated into the process flow for fabricating flexible capacitive pressure sensors. With this approach, the prefabricated molds are not needed. As illustrated in Fig. 6a, the mixed solution of PDMS prepolymer and its curing agent with ammonium bicarbonate (NH_4HCO_3) were drop-casted onto an ITO-coated PET film. NH_4HCO_3 is a commonly used foaming agent in the food industry. The other ITO/PET film was laminated on top with two thin glass strips as the spacers, defining the gap distance between two electrodes and in turn the thickness of the PDMS dielectric layer. Then, the sample was annealed at 100 °C for 4 h to cure the PDMS prepolymer, and NH_4HCO_3 was decomposed into NH_3, H_2O, and CO_2 at the same time, forming microstructured PDMS film of high-density air void micro-features as shown in Fig. 6b. With such a facile approach, the sensor performance can be optimized through varying the NH_4HCO_3 concentration and the PDMS film thickness. The fabricated large area samples can be cut into small pieces of sensor devices for use.

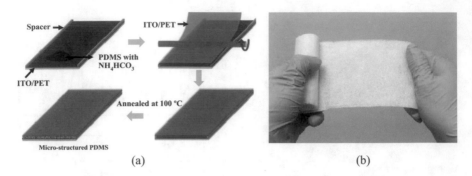

(a)   (b)

Fig. 6 (a) Schematic illustration of one-step processing of the microstructured PDMS film based on a mixture of PDMS prepolymer and its curing agent with ammonium bicarbonate (NH_4HCO_3) and its seamless integration into the process flow for fabricating a flexible capacitive sensor. (b) The photo image of the fabricated large area microstructured PDMS film. (Edited from [22])

When an external pressure is applied, the PDMS film can be easily compressed with the deformation of air voids (Fig. 7a). As shown in Fig. 7b, the sensor device made of this microstructured PDMS film has a sensitivity of 0.26 kPa^{-1} in the low-pressure regime (0–0.33 kPa), which is close to that using prefabricated pyramid structure PDMS layer via complicated microfabrication processes (0.55 kPa^{-1}) [10]. Figure 7c shows the sensing responses of the device upon placement and removal of loads in low-, medium-, and high-pressure ranges. It can be seen that the sensor can be used to detect loads over a wide pressure range with fast response time of less than 15 ms. Compared with previous works on capacitive sensors using elastic dielectric layer [13–18], a significant advantage of this sensor is the capability of maintaining a relatively high sensitivity in a much wider pressure range, which enables to detect small pressure changes at heavy loads.

The endurance test for the sensor device was carried out by hitting the device with a hammer, which induced a pressure load exceeding 1 MPa, as shown in Fig. 7d. The sensor can recover to its initial value quickly after each hit, which further proves the excellent endurance of the device.

With high sensitivity, fast response, and excellent endurance over a large pressure range, the sensor can thus be used to monitor a wide range of human activities at different pressure levels from wrist pulses to foot plantar pressure [23, 24]. Here pressure sensors were fabricated and integrated in insoles, which allow the users to accurately monitor their walking steps and speed and assess the energy expenditure for the promotion of healthy lifestyle.

Moreover, static and dynamic measurement of foot pressure and its distribution across the plantar foot surface and bony structures is clinically essential to identify anatomical foot deformities and guide the diagnosis and treatment of gait disorders and leads to strategies for identifying significant foot-related sport injury risk such as ankle sprain and preventing pressure ulcers in diabetes [25]. Real-time foot pressure measurement is also helpful to guide the postsurgical patients to resume a correct load distribution over both legs in order to facilitate bone osteogenesis [26].

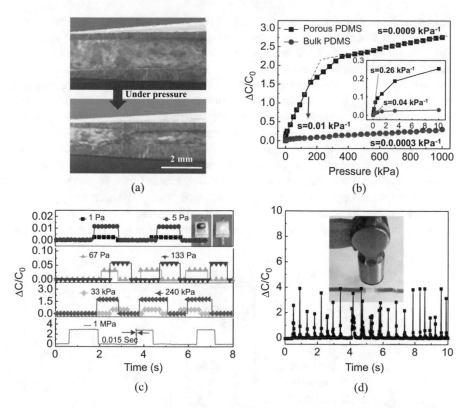

(a)   (b)

(c)   (d)

Fig. 7 (**a**) The cross-sectional photo images of the microstructured PDMS film clipped by a tweezer without pressure (upper) and with pressure (bottom). (**b**) The measured capacitance changes as a function of the applied pressure for the pressure sensors using the microstructured and unstructured PDMS films (inset: the display of the results in the low pressure range (0–10 kPa)). (**c**) The transient response of the pressure sensors upon the applied pressure at different pressure levels (inset: a red bean (75 mg) and a rice grain (15 mg), respectively, on a thin glass slide (15 mm × 10 mm, 150 mg), which covers the entire sensor area, for measurement in the low-pressure range). (**d**) The durability of the pressure sensor hit by a hammer with pressure loads larger than 1 MPa. (Edited from [22])

For these smart insole applications, in addition to the basic requirements of low cost and maintaining comfort of wear, the sensors need to sustain high-pressure loads with high durability while being able to detect small pressure changes. The exhibited excellent features of high sensitivity over a wide pressure range, excellent durability under heavy loads, and very simple fabrication processes enable such a microstructured PDMS film-based sensor to be an ideal technology of choice for smart insole applications.

As shown in Fig. 8a, seven sensors were integrated in the fabricated insole, which were placed at the key locations of the foot area according to gait kinetics as well as normal and pathological foot anatomy [22]. Although more sensor points can be easily made on the insole with this sensor technology, a reduced number of sensors will be helpful to reduce the cost of the control electronics and the power consumption. The outputs of the seven sensors were connected to the self-designed DAQ circuit board with 14 measurement channels, which communicated with the mobile phone through a Bluetooth module. A continuous image of foot plantar pressure can be reconstructed from the data of a limited number of sensors for gait analysis (Fig. 8b). The shoe with the insole inside was worn by a volunteer for tests. The high sensitivity under high-pressure loads with this sensor also enables the worn smart insole to detect the change less than 1 kg (corresponding to a pressure change of 0.3 kPa) for a 70 kg adult as shown in Fig. 8c. When he walked or run, the dynamic capacitance changes can be real time monitored through the sensors as shown in Fig. 8d. From the measured data on one sensor (point 6), the walking or running status can be clearly clarified, and the steps can also be accurately counted. The transient results of the seven sensors under walking and running conditions are given in Fig. 8e, f. The results prove that such a smart insole system can be used for

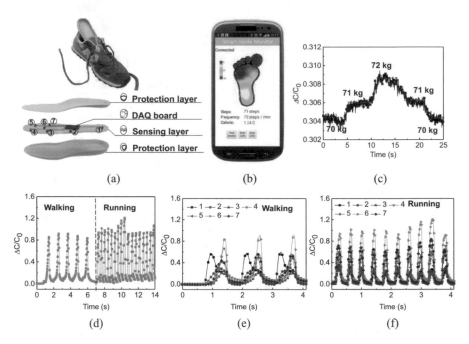

Fig. 8 (**a**) Illustration of the structure of the smart insole with seven sensors being connected to a self-designed DAQ circuit board, which communicates with the mobile phone through a Bluetooth module. (**b**) A continuous image of foot plantar pressure is reconstructed from the data of the seven sensors for gait analysis, which is processed in the mobile phone with a software developed in Android. (**c**) The measurement results show the capability of the smart insole for detecting very small weight change of less than 1 kg (corresponding to a pressure change of 0.3 kPa) for a 70 kg adult. (**d**) The measured real-time dynamic capacitance changes of one sensor (point 6) clearly show the walking and running conditions and the walked or run steps. (**e**) The transient results of the seven sensors under walking. (**f**) The transient results of the seven sensors under running. (Edited from [22])

gait analysis to provide key clinical and sports information. With the demonstrated capabilities of real-time monitoring the walking or running speed and steps and measuring the slight weight change, the smart insole based on the developed sensor would be a promising technology to be used for evaluation of sports performance during training or competition and assessment of energy expenditure in daily life.

4   Ultrathin PDMS Films for Strain Sensors

A general approach to obtain PDMS films for sensor devices is coating the mixed solution of PDMS prepolymer and its curing agent on a supporting substrate such as a glass slide or a silicon wafer and then peeling off the PDMS film from the substrate after thermal curing. To easily peel off the PDMS film from the substrate without damages, a PDMS film with thickness of a few hundred micrometers (>500 μm) is generally required to sustain the tensile stress during the process [27, 28], which is determined by the adhesion strength between the PDMS film and the supporting substrate. Such a thick PDMS film is difficult for fabricating sensors to meet the unperceivable and invisible wearable application requirements.

The adhesion strength between the PDMS and the glass substrate was shown to be effectively decreased by surface treatment of the glass substrate with a fluoroalkylsilane-trichloro (1H,1H,2H,2H-perfluorooctyl)silane (FOTS) self-assembled monolayer (SAM) to reduce the surface wetting (Fig. 9a). A very thin PDMS film was thus able to be easily fabricated. To fabricate strain sensors, silver nanowires (AgNWs) were used to be incorporated into the PDMS film considering their advantages of high dc conductivity and optical transmittance, excellent mechanical flexibility, and environmental and economical advantages for manufacturing [29]. The fabrication process of the ultrathin AgNW/PDMS

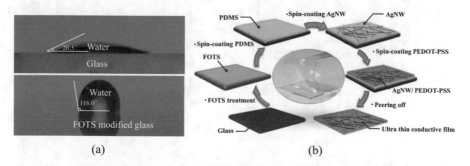

(a)   (b)

Fig. 9 (a) The measured water contact angles on glass substrate before and after surface treatment with the fluoroalkylsilane-trichloro (1H,1H,2H,2H-perfluorooctyl)silane (FOTS) self-assembled monolayer (SAM). (b) Illustration of the fabrication procedure for the ultrathin (~20 μm) poly(dimethylsiloxane) (PDMS) elastomer film integrating a conductive composite of silver nanowires (AgNWs) and neutral-pH poly(3,4-ethylenedioxythiophene):poly(styrenesulfonate) (PEDOT:PSS) for strain sensors. (Edited from [19])

composite film is depicted in Fig. 9b. The surface of a clean glass slide (4 cm × 4 cm) was treated with FOTS in a vacuum. With a hydrophobic substrate surface after the FOTS treatment to decrease the adhesion strength between the PDMS and the substrate, an ultrathin (about 20 μm) AgNW/PDMS composite film was able to be easily peeled off from the glass substrate, as illustrated in the center of Fig. 9b.

The fabricated 20 μm-thick PDMS film presents nearly 80% transparency over the visible light wavelength range from 400 to 800 nm, while the 800 μm-thick one having much poorer transparency less than 60% (Fig. 10a). The results indicate the advantage of using ultrathin elastomer film for transparent invisible wearable sensors. Figure 10b shows that with an applied tensile force of 0.196 N (20 g counterweight) to the 20 μm-thick film sensor, the resulted mechanical strain (ε)

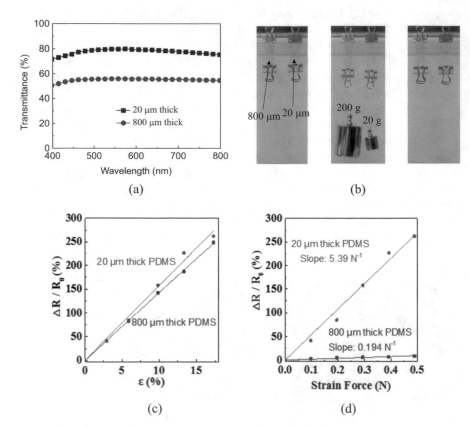

(a)   (b)

(c)   (d)

Fig. 10 (**a**) The measured optical transmittance of the strain sensors with 20 μm and 800 μm-thick PDMS films, respectively. (**b**) Photo images comparing the mechanical deformation upon applying and removing tensile force for the sensors with different thick PDMS films (800 μm and 20 μm). (**c**) The measured related changes in the electrical resistance ($\Delta R/R_0$) at different mechanical strains (ε) for both devices. (**d**) The measured $\Delta R/R_0$ at different applied strain forces for both devices. (Edited from [19])

of about 10% (defined as the relative length change $\Delta L/L$) is similar to that of the 800 μm-thick sensor device with a tensile force of 1.96 N (200 g counterweight). The ultrathin film sensor is much easier to be deformed under tensile force. After removal of the loads, both sensors recovered to their initial states. For wearable applications, the required tensile force (F) to induce a certain mechanical strain should be minimized to make the sensors unperceivable to the users. According to Hooke's law [30], the tensile force can be expressed as:

$$F = \varepsilon \cdot E \cdot w \cdot d \tag{1}$$

where E is the Young's modulus and w and d are the width and the thickness of the strain sensor, respectively. Therefore, to achieve the same mechanical strain (determined by the human activities to be monitored), less tensile force is required on the device of a thinner PDMS film. The measured related changes in the electrical resistance ($\Delta R/R$) at different mechanical strains (ε) for both devices are shown in Fig. 10c, presenting the similar gauge factor (GF) of about 15. The gauge factor (GF) is defined as the ratio of $\Delta R/R$ to ε [31]. With the similar GF, but much larger mechanical strain being induced under the same tensile force, the 20 μm-thick film sensor presents a tensile force sensitivity (S) of 5.39 N^{-1}, which is more than 20 times higher than that of the 800 μm-thick film device (0.194 N^{-1}) as shown in Fig. 10d. S is defined as the ratio of $\Delta R/R$ to the applied tensile force (F) in the form of $S = (\Delta R/R_0)/F$. For the ultrathin film sensor with much improved S, the required tensile force for a certain mechanical strain is significantly minimized, which has benefits to achieve unperceivable sensors for wearable applications.

The long-term operation stability of the ultrathin PDMS film sensor was carried out by applying 1000 cycles of repeated stretching/releasing with a maximum strain of 20% to the sensor for 1400 s based on the experimental setup shown in Fig. 11a. The results in Fig. 11b indicate that the ultrathin film strain sensor can have excellent mechanical durability against long-term repeated elongation/relaxation cycles for monitoring human motion. Finally, the fabricated ultrathin PDMS film sensor was demonstrated for neck posture monitoring. Very large percent of the population suffers with neck pain and its associated problems caused by prolonged poor neck posture during work or life (e.g., when texting, reading, gaming, or watching media information with a computer, a smartphone, or other devices) [32]. When the head is bent forward and down, the neck posture becomes away from the natural curve of the cervical spine, and the stress on the cervical spine will be substantial depending on the angle of the bend and the amount of time spent with this neck posture. The stress may lead to early wear and tear on the spine, early spinal degeneration, muscle strain, pinched nerves, herniated discs, and abnormalities to the neck's natural curvature and has also been linked to headaches, neurological problems, and heart diseases [32]. To help individuals rectify their neck posture timely to avoid or minimize these issues, real-time monitoring of neck posture is very essential. In this work, the ultrathin film sensor was stuck to a volunteer's neck with medical adhesive tape. The output of the sensor is connected to the self-designed DAQ circuit board, which is fixed onto the collar of the coat. The circuit board received the sensor

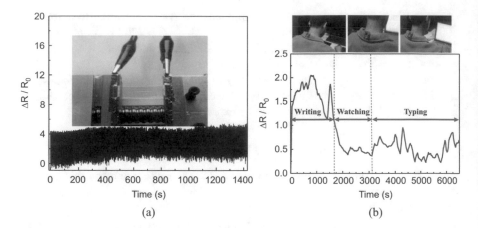

Fig. 11 (a) The durability test results by applying 1000 cycles of repeated stretching/releasing with a maximum strain of 20% to the strain sensor for 1400 s. Inset: photo image of the experimental setup for durability test for the strain sensor. (b) Real-time monitoring of a volunteer's neck posture with the fabricated thin PDMS film strain sensor, when the volunteer performed normal activities in the working place including writing, watching the computer monitor, and typing. (Edited from [19])

signals and communicated wirelessly with the mobile phone through Bluetooth. A software developed in Android was deployed in a mobile phone to display the data on the screen in real time. The volunteer's neck posture was continuously monitored for about 2 h when he performed normal activities in the working place including writing, watching the computer monitor, and typing, as shown in Fig. 11b. The results prove that the sensor can be used for neck posture real-time monitoring. Further, with its features of high sensitivity (low tensile force for sensing) and high transparency, the ultrathin film technology would be an ideal choice for developing invisible and unperceivable sensors in wearable applications.

5   Conclusions

Polydimethylsiloxane (PDMS) elastomer has been popularly used as base materials for constructing various kinds of pressure and strain sensors, due to its excellent flexible and elastic properties, biomedical compliance with human tissues, and commercial availability. For high sensitivity and imperceptible sensors to be worn on human body, it is important to obtain microstructured and ultrathin PDMS films. Engineering techniques with low-cost scalable manufacturing processes are developed for this purpose. Typical application examples, including wrist pulse monitoring, static and dynamic foot pressure monitoring, and neck posture monitoring, were demonstrated. These results prove such engineering of PDMS elastomer would be able to provide a versatile technology platform for developing various pressure and strain sensors to meet the wearable health applications.

References

1. Zhu H, Wang X, Liang J et al (2017) Versatile electronic skins for motion detection of joints enabled by aligned few-walled carbon nanotubes in flexible polymer composites. Adv Funct Mater 27:1606604
2. Gerratt AP, Michaud HO, Lacour SP (2015) Elastomeric electronic skin for prosthetic tactile sensation. Adv Funct Mater 25:2287–2295
3. Pang Y, Tian H, Tao L et al (2016) Flexible, highly sensitive, and wearable pressure and strain sensors with graphene porous network structure. ACS Appl Mater Interfaces 8:26458–26462
4. Wang C, Li X, Gao E et al (2016) Carbonized silk fabric for ultrastretchable, highly sensitive, and wearable strain sensors. Adv Mater 28:6640–6648
5. Lee D, Lee H, Jeong Y et al (2016) Highly sensitive, transparent, and durable pressure sensors based on sea-urchin shaped metal nanoparticles. Adv Mater 28:9364–9369
6. Li T, Luo H, Qin L et al (2016) Flexible capacitive tactile sensor based on micropatterned dielectric layer. Small 12:5042–5048
7. Jeong SH, Zhang S, Hjort K et al (2016) PDMS-based elastomer tuned soft, stretchable, and sticky for epidermal electronics. Adv Mater 28:5830–5836
8. Joo Y, Byun J, Seong N et al (2015) Silver nanowire-embedded PDMS with a multiscale structure for a highly sensitive and robust flexible pressure sensor. Nanoscale 7:6208–6215
9. Zhao X, Hua Q, Yu R et al (2015) Flexible, stretchable and wearable multifunctional sensor array as artificial electronic skin for static and dynamic strain mapping. Adv Electron Mater 1:1500142
10. Mannsfeld SC, Tee BC, Stoltenberg RM et al (2010) Highly sensitive flexible pressure sensors with microstructured rubber dielectric layers. Nat Mater 9:859–864
11. Park H, Kim DS, Hong SY et al (2017) A skin-integrated transparent and stretchable strain sensor with interactive color-changing electrochromic displays. Nanoscale 9:7631–7640
12. Madsen FB, Daugaard AE, Hvilsted S et al (2016) The current state of silicone-based dielectric elastomer transducers. Macromol Rapid Commun 37:378–413
13. Park J, Lee Y, Hong J et al (2014) Giant tunneling piezoresistance of composite elastomers with interlocked microdome arrays for ultrasensitive and multimodal electronic skins. ACS Nano 8:4689–4697
14. Pang C, Koo JH, Nguyen A et al (2015) Highly skin-conformal microhairy sensor for pulse signal amplification. Adv Mater 27:634–640
15. Boutry CM, Nguyen A, Lawal QO et al (2015) A sensitive and biodegradable pressure sensor array for cardiovascular monitoring. Adv Mater 27:6954–6961
16. Tee BCK, Chortos A, Dunn RR et al (2014) Tunable flexible pressure sensors using microstructured elastomer geometries for intuitive electronics. Adv Funct Mater 24:5427–5434
17. Kang S, Lee J, Lee S et al (2016) Highly sensitive pressure sensor based on bioinspired porous structure for real-time tactile sensing. Adv Electron Mater 2:1600356
18. Lee BY, Kim J, Kim H et al (2016) Low-cost flexible pressure sensor based on dielectric elastomer film with micro-pores. Sensors Actuators A 240:103–109
19. Hu W, Chen S, Zhuo B et al (2016) Highly sensitive and transparent strain sensor based on thin elastomer film. IEEE Electron Device 37:667–670
20. Xu F, Zhu Y (2012) Highly conductive and stretchable silver nanowire conductors. Adv Mater 24:5117–5122
21. Zhuo B, Chen S, Zhao M et al (2017) High sensitivity flexible capacitive pressure sensor using polydimethylsiloxane elastomer dielectric layer micro-structured by three-dimensional printed mold. IEEE J Electron Devices 5:219–223
22. Chen S, Zhuo B, Guo X (2016) Large area one-step facile processing of microstructured elastomeric dielectric film for high sensitivity and durable sensing over wide pressure range. ACS Appl Mater Interfaces 8:20364–20370
23. Wang X, Gu Y, Xiong Z et al (2014) Silk-molded flexible, ultrasensitive, and highly stable electronic skin for monitoring human physiological signals. Adv Mater 26:1336–1342

24. Razak AHA, Zayegh A, Begg RK et al (2012) Foot plantar pressure measurement system: a review. Sensors 12:9884–9912
25. Paton J, Bruce G, Jones R et al (2011) Effectiveness of insoles used for the prevention of ulceration in the neuropathic diabetic foot: a systematic review. J Diabetes Complications 25:52–62
26. Jarchi D, Lo B, Ieong E, et al. (2014) Validation of the e-AR sensor for gait event detection using the parotec foot insole with application to post-operative recovery monitoring. In: Proceedings of international conference on wearable & implantable body sensor networks, Zurich, Switzerland, pp 127–131
27. Wang Y, Wang L, Yang T et al (2014) Wearable and highly sensitive graphene strain sensors for human motion monitoring. Adv Funct Mater 24:4666–4670
28. Yan C, Wang J, Kang W et al (2014) Highly stretchable piezoresistive graphene-nanocellulose nanopaper for strain sensors. Adv Mater 26:2022–2027
29. Hu L, Kim HS, Lee JY et al (2010) Scalable coating and properties of transparent, flexible, silver nanowire electrodes. ACS Nano 4:2955–2963
30. Altenbach H, Altenbach JW, Kissing W (2004) Mechanics of composite structural elements. Springer, Berlin, pp 18–19
31. Hempel M, Nezich D, Kong J et al (2012) A novel class of strain gauges based on layered percolative films of 2D materials. Nano Lett 12:5714–5718
32. Hansraj KK (2014) Assessment of stresses in the cervical spine caused by posture and position of the head. Surg Technol Int 25:277–279

Write Mode Aware Loop Tiling for High-Performance Low-Power Volatile PCM in Embedded Systems

Keni Qiu, Qingan Li, Jingtong Hu, Weigong Zhang, and Chun Jason Xue

1 Introduction

DRAM-based main memory is facing huge challenges due to the scalability limitation as well as the energy efficiency problem. As an alternative, phase change memory (PCM) is becoming a promising replacement to be deployed as main memory in the deep submicron regime [1–3]. PCM has the advantages of comparable read speed as DRAM, near zero leakage power consumption, good scalability, and nonvolatility [4]. However, PCM suffers long latency and high energy on write operations [5–7]. Multiple level cell (MLC) PCM pays the cost of slow write performance while increasing the density of PCM [8, 9]. With a limited power budget, slow and high energy PCM writes can hurt system performance [10]. Loops are usually the most computation-intensive part of embedded applications. This work aims to optimize write performance and energy for loops on MLC PCM.

K. Qiu (✉) · W. Zhang
College of Information Engineering at Capital Normal University, Beijing, China
e-mail: qiukn@cnu.edu.cn; zwg771@cnu.edu.cn

Q. Li
State Key Laboratory of Software Engineering at Wuhan University, Wuhan, China
e-mail: qingan@whu.edu.cn

J. Hu
Department of Electrical and Computer Engineering, Swanson School of Engineering, University of Pittsburgh, Pittsburgh, PA, USA
e-mail: jthu@pitt.edu

C. J. Xue
Department of Computer Science at City University of Hong Kong, Kowloon Tong, Hong Kong
e-mail: jasonxue@cityu.edu.hk

© Springer Nature Switzerland AG 2020
Y. Liu et al. (eds.), *Smart Sensors and Systems*,
https://doi.org/10.1007/978-3-030-42234-9_10

171

A lot of exciting work has been done to reduce energy consumption for MLC PCM [11, 12]. Recent work shows the potential of trading off write performance with data retention time for MLC PCM with write speed option. The research in [13, 14] shows that the nonvolatility of MLC PCM can be traded for better performance and lower write energy. By architecting MLC PCM as the main memory for microcontroller units (MCUs), Li et al. have proposed a compiler directed dual-write (CDDW) scheme to adopt two types of write modes, referred as *slow* mode and *fast* mode, for different write operations [15]. With respect to the hardware limitation in MCU systems which do not consider memory refreshes, the CDDW scheme can totally avoid refresh operations. Significant improvements on performance and write energy by the CDDW scheme are reported for the selected MiBench programs [15].

This work focuses on loops with intensive data array operations, which have read and write dependencies. This class of loops is often used to implement relaxation methods in numerical simulations and signal processing [16]. We observe that, as the loop scales up, the lifetime of write instances will be very long, necessitating slow write mode. In this scenario, the CDDW scheme exhibits very low efficiency in performance and dynamic energy improvement.

Loop tiling is a classic loop transformation technique to enhance data locality [16–20]. In this work, we employ the loop tiling technique for a novel objective. We propose to effectively implement loop tiling to reduce the lifetime of write instances in loop nests. With loop tiling, the define-use chains of most data write instances in each tile are limited within the tile so that their lifetimes are also limited within the tile's execution time. With dedicated tile size selection, the new lifetime of write instances can be reduced below the fast write retention time. Consequently, most original slow write instances can be written in fast mode. This proposed write mode aware loop tiling approach, incorporated with the CDDW scheme, can achieve high performance and low power for loops on MLC PCM. We conducted experiments to evaluate the proposed scheme for different loop kernels. Results show that the proposed loop tiling approach improves the performance by 70.0%/51.8%/50.8% and reduces dynamic energy by 35.6%/33.7%/32.0% compared to the fast/slow/CDDW scheme on average.

The rest of this work is organized as follows. The volatile MLC PCM model and the CDDW scheme are introduced in Sect. 2. A motivational example is presented to illustrate the benefit of loop tiling in Sect. 3. The proposed write mode aware loop tiling approach is described in detail in Sect. 4. Section 5 presents the evaluation results. Finally, we conclude the work in Sect. 6.

2   Background

In this section, we introduce the volatile PCM model and the CDDW scheme.

2.1 Volatile PCM Model

We first introduce the study background on MLC PCM write operations and the trade-off between MLC PCM write latency (energy) and the retention time.

2.1.1 MLC PCM and its Write

PCM utilizes the phase-change behavior of chalcogenide glass such as GST ($Ge_2Sb_2Te_5$) to record data. By injecting electrical pulses to Joule heater (Fig. 1a), GST can be switched between large resistance state (amorphous state) and small resistance state (crystalline state). MLC PCM exploits intermediate resistance levels between these two states to store multiple logic bits per cell.

Due to process variations and material composition fluctuation, different PCM cells in one memory line respond distinctively to programming pulses, and even the same cell responds differently at different time. Therefore, PCM widely adopts an iteration-based programming and verifying (P&V) write scheme (Fig. 1b) to precisely control the cell resistance. A *RESET* operation is always first conducted to put the cell in an initial state. A series of *SET* and verify (read) operations then follow until the target resistance level is reached.

2.1.2 Trade-Off Between Write Latency and Retention Time

The multilevel states of resistances of MLC PCM follow a normal distribution considering process variation [21–24]. As Fig. 2 shows, MLC PCM uses resistance ranges to represent the information stored in the cell. Four resistance ranges indicate four data value, from "01" to "00" (gray encoding). For MLC PCM, a small resistance range, referred as *guardband*, is intentionally left between two resistance states to prevent the lower resistance state from drifting into the higher resistance state [25]. Due to the relaxation of the parameters of amorphous phase,

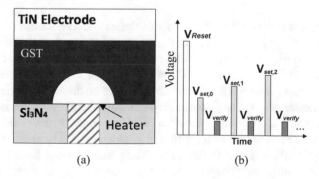

(a)   (b)

Fig. 1 PCM cell and its write operation. (**a**) PCM cell. (**b**) MLC PCM write

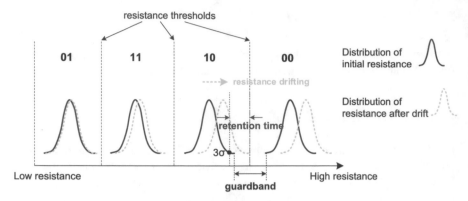

Fig. 2 MLC PCM resistance distribution and resistance drift

the resistance of PCM spontaneously increases [13]. This phenomenon is known as *resistance drift*. In Fig. 2, the broken curves illustrate the resistance distribution after drift. And the drifting speed is proportional to the volume fraction of amorphous phase [13]. Drift affects the stability of the electrical behavior of MLC PCM cell and thus the reliability of MLC storage.

In [13], the guardband between "10" and "00" is identified as the most drift-prone one. We use the guardband between "10" and "00" as illustrated in Fig. 2 as illustration. Longer retention time can be obtained by increasing guardband size, since larger guardband can tolerate larger resistance drifting. Large guardband requires tight distribution of each neighborhood resistance state. To increase the programming accuracy and achieve tight resistance distribution, more write iterations and energy must be paid in each write operation. MLC PCM cells, if written according to this setting, are considered as *nonvolatile*. On the contrary, shorter write latency produces smaller guardband, which introduces smaller retention time. In this case, the cells are considered as *volatile*.

2.1.3 Volatile MLC PCM Model

The current/resistance of MLC PCM model calculates the resistance of MLC PCM cell when applying a RESET or SET current on the cell for a given latency [26]. MLC PCM usually adopts iteration-based programming method. By applying different amplitudes of current on PCM cell [15], different resistances can be obtained. A process variation distribution is produced on the key parameters of PCM cell, such as heater radius. The distance between worst-case resistances is selected as the PCM resistance guardbands. Using the PCM resistance drifting model proposed in [13], different retention times for different guardbands can be obtained [26]. This retention time is defined as the shortest drifting time elapsed from the time that the resistance is only just recognizable to the time that the resistance reaches the corresponding state threshold. We consider the worst-case drifting distance (starting

Table 1 The volatile MLC PCM model

Iteration (#)	Current (μA)	N. energy	Retention time (s)
10	310	1 (baseline)	11158.84
8	320	0.85	4823.178
7	330	0.75	2084.719
6	340	0.72	713.7916
5	360	0.674	83.67949
4	380	0.6	20.67646
3	410	0.524	1.87

from the 3σ point in the normal distribution and ending on the middle point of the guardband between states "10" and "00") and worst-case drifting state (the most drift-prone state "10") when calculating the retention time. In this way, the derived retention time covers the worst cases and thus provides safe boundary of retention time for volatile storage. The volatile MLC PCM model is summarized in Table 1 [15]. This MLC PCM model supports multilevel retention time and write latencies.

To ensure data reliability, in general, the volatile write mode (or fast mode) needs periodical refresh if the write instances' lifetime exceeds the retention time, which not only involves hardware overhead but also consumes additional energy and execution cycles. On the other hand, the nonvolatile mode (or slow mode) provides long enough lifetime insurance, but leads to longer write latency and energy inefficiency.

2.2 CDDW Scheme

MCUs are commonly applied in embedded systems. From the perspective of the hardware limitation in MCU systems, we do not consider dedicated refresh hardware support in the MCU design. Targeting the computing system consisting of MCU- and PCM-based main memory with no caches, Li et al. [15] propose the CDDW approach to exploit the trade-off between performance and retention time of PCM. In the CDDW scheme, different write modes are selected for different memory write instructions (MWIs) according to their lifetime. The lifetime of a MWI instance is defined as the elapsed time from the time that this MWI instance writes a value into a memory line to the time that the last read of this value occurs. Each MWI instance starts a new lifetime. The lifetime of a MWI is illustrated in Fig. 3.

The characteristics of PCM writes demonstrate that, if the lifetime of a MWI instance is shorter than the retention time of PCM cells, no refresh is needed for this write and the data correctness can be guaranteed. In summary, the dual write mode in the CDDW scheme involves two aspects. First, for a MWI with lifetime longer than the fast write retention time, slow mode is applied to it to avoid refreshes. And

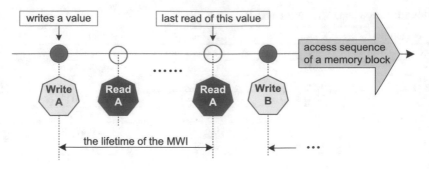

Fig. 3 The lifetime of a MWI

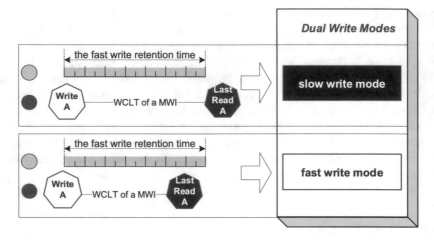

Fig. 4 The dual write modes in the CDDW scheme

second, for a MWI with lifetime shorter than the fast write retention time, fast mode is applied to it to enjoy short latency and energy saving for write operations.

It is not practical to identify the exact lifetime of each MWI instance at runtime, since such kind of information is sensitive to program structure and program inputs. The CDDW utilizes the worst-case lifetime (WCLT) of each MWI to represent the MWI's lifetime based on static analysis. The CDDW scheme is illustrated in Fig. 4. More details can be found in [15].

3   Motivation

With respect to large-scale loops, we observe that the CDDW scheme is often fruitless because the lifetimes of most MWIs are too long, necessitating the expensive slow mode. Figure 5a shows a nested loop example. Figure 5b shows

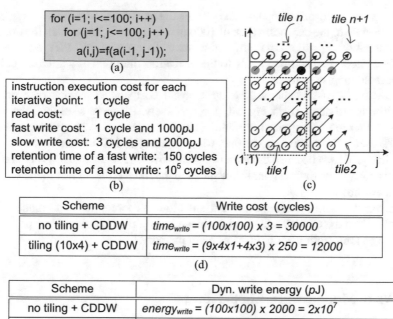

Fig. 5 Motivational example. (**a**) A loop nest. (**b**) Architectural parameters. (**c**) Loop tiling in the iteration space. (**d**) Write cost comparisons under CDDW with tiling and with no tiling. (**e**) Dynamic write energy comparisons under CDDW with tiling and with no tiling

the assumptions of architectural parameters. There exist read after write (RAW) dependencies in the data array computations as depicted in Fig. 5c. With no loop tiling, each MWI's lifetime is constant and equals to the execution time of 101 iteration points. That is, each MWI's lifetime is 303 ($101 \times (1 + 1 + 1)$) cycles using fast mode (without counting refresh cost) or 505 ($101 \times (3 + 1 + 1)$) cycles using slow mode. Since the fast write retention time is only 150 cycles, all the write instances should adopt slow mode according to the dual write mode selection metric in the CDDW scheme [15].

We find that the loop tiling technique can be utilized to reduce MWIs' lifetime and thus improve the MLC PCM's access performance and save write energy. We tentatively partition the iteration space of the loop nests into small tiles with size of 10×4 as indicated by red boxes in Fig. 5c and then execute them one by one according to the default row-wise direction. It can be seen that the lifetime of each MWI in the dashed rectangle is less than or equal to the execution time of 45 iteration points for each tile. In other words, the lifetime of MWIs in this area is 135 cycles using fast mode or 225 cycles using slow mode. Since lifetimes of MWIs in the dashed rectangle are shorter than the fast write retention time, we can select fast mode for these 36 (9×4) writes according to the CDDW scheme. In terms of the MWIs occurring at the top boundary of the tile as indicated by gray circles, their lifetimes equal to the execution time of 1000 iteration points, larger than the original ones without tiling, still necessitating slow mode. In terms of the MWI indicated by the black circle, its lifetime equals to the execution time of 1041 iteration points, also necessitating slow mode.

Under the CDDW scheme, the write cycle cost and dynamic write energy comparisons are depicted in Fig. 5d, e, respectively. It can be seen that, the write cycles and write energy with loop tiling (12,000 cycles/1.1×10^7 pJ) are reduced by 70% and 45% comparing to that with no tiling (40,000 cycles/2×10^7 pJ) respectively. This is because the loop tiling effectively reduces most MWIs lifetime and thus transforms their original slow mode to the more efficient fast mode with no refresh overhead.

Motivated by the example, this work aims to incorporate the write mode aware loop tiling into the CDDW scheme, to achieve high performance and low power for loops when architecting MLC PCM as main memory.

4 Write Mode Aware Loop Tiling

Tile shape and tile size are the key factors when implementing loop tiling transformations. The tile shape should guarantee legal tiling. The tile size can impact how many MWIs achieve lifetime reduction and how much their lifetime is reduced. In this section, we discuss how to determine the legal tile shape and the optimal tile size.

4.1 Overview

Data array computations in loops lead to the characteristic of constant dependency vectors, i.e., data dependencies that have a constant distance in the iteration space. As loops scale up, MWIs' lifetime will exceed fast write retention time, necessitating slow mode. This work proposes a write mode aware loop tiling (WMALT) approach to apply the optimal loop tiling to loop nests and effectively enable maximal fast writes for performance and energy improvements.

The proposed WMALT approach consists of two parts: legal tile shape determination and optimal tile size selection. The first part determines the directions of tiling vectors to guarantee legality. The second part calculates the optimal tile size to effectively reduce MWIs' lifetime so as to maximize fast writes among the MWIs.

For simplicity, we use two-level loop nests without instructions between loops as examples to derive the optimal loop tiling solution. The two dimensions are generically referred as y and x dimensions. The solution of the multilevel case can be derived analogously.

4.2 Legal Tile Shape

Loop transformation must preserve the temporal sequence of all dependencies. In order to do legal tiling, we first need to analyze the data dependencies in loops. There are totally four kinds of data dependencies: read after write (RAW, flow dependence), write after read (WAR, antidependence), write after write (WAW, output dependence), and read after read (RAR, input dependence). The RAW and WAR dependencies prohibit execution sequence reordering. This work focuses on the MWI's lifetime analysis which is correlated to the RAW dependency.

A dependency vector $d = (d_y, d_x)$ means the computation of a data array at iteration (i, j) depends on the execution of the data array at iteration $(i - d_y, j - d_x)$. In the example shown in Fig. 6, it is assumed there are four dependency vectors: d_1, d_2, d_3, and d_4 in two-level nested loops, where d_3 is the extreme CCW (counterclockwise) vector and d_4 is the extreme CW (clockwise) vector. Legal tile vectors can only be outside of CW and CCW or aligned with them [27]. In Fig. 6, we choose the tile vector T_y to be aligned with CCW and the tile vector T_x with x-axis, which is outside of CW. It is a legal tiling direction choice as indicated by the dotted lines. We observe that the y-elements of dependency vectors are mostly positive or zero in the nested loops for real applications. Therefore, we choose the T_y vector to be aligned with the CCW vector and T_x vector to be aligned with the positive x-axis; thus a legal tiling can be obtained.

It is not always possible to do legal tiling directly. The work [19] points out that a loop i through a loop j in nested loops can realize tiling if and only if they are fully permutable. In general, we can use loop transformation techniques such as loop retiming, skewing, etc. to preprocess the loops and then obtain legal tiling.

4.3 Optimal Tile Size

In previous studies, loop tiling has often been employed to tailor loop nests into smaller tiles to improve cache hit rate. In this work, we use loop tiling for the novel

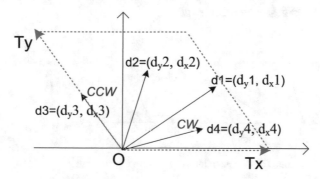

Fig. 6 Legal tile shape determination

purpose of reducing MWI's lifetime and enable as many as possible write operations to be written in fast mode. Hence, the overall efficiency of the MLC PCM can be improved.

The definition of the lifetime of a MWI implies that MWI's lifetime analysis is only correlated to the RAW dependency. Therefore, we only consider the RAW dependencies and the CCW dependency vector when determining the tile size.

For easier analysis, two constraints are imposed on the tiling vectors T_y and T_x:

- T_y **Constraint:** The vectors T_y shall be large enough so that no RAW dependency vector starting from the tile origin can pass through an entire tile along the y-element directions.

- T_x **Constraint:** The vectors T_x shall be large enough so that no RAW dependency vector starting from the tile origin can pass through an entire tile along the x-element directions.

Figure 7a, b presents two cases that are against the T_y constraint and the T_x constraint, respectively. In Fig. 7a, the tiling vector along the y-element direction is not large enough, because the RAW dependency vector d_1 crosses the top boundary of the tile. In Fig. 7b, the tiling vector along the x-element direction is not large enough, because the RAW dependency vector d_2 crosses the right boundary of the tile.

Before deriving the optimal tile size, we first propose the concept of *base tile*. A base tile is a tiling solution, satisfying the above constraints and having the minimal tile lengths along the y-element and x-element directions. The T_y and T_x constraints guarantee that all the RAW dependency vectors that are starting from the origin of a base tile should be included within the base tile.

A base tile is expressed as $T_y^{\text{base}} \times T_x^{\text{base}}$ where T_y^{base} and T_x^{base} denote its y-element and x-element tile lengths, respectively. It is assumed that there are three dependency vectors, $d_1 = (d_{y1}, d_{x1})$, $d_2 = (d_{y2}, d_{x2})$, and $d_3 = (d_{y3}, d_{x3})$, denoting the ones with the largest x-element dependency length, the largest y-element dependency length, and the CCW direction in loop nests, respectively, as shown in Fig. 7c. In particular, we use $d_y\text{max}$ and $d_x\text{max}$ to represent the largest y-element length and x-element length, respectively. In the case of Fig. 7c, we have $d_y\text{max} = d_{y2}$, $d_x\text{max} = d_{x1}$. It is further assumed that α denotes the angle of the dependency vector in CCW direction and β denotes the angle of the dependency vector with the largest x-element length.

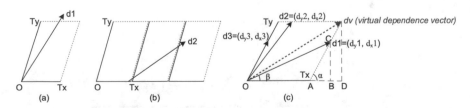

Fig. 7 (a) T_y constraint. (b) T_x constraint. (c) The base tile calculation

In order to cover all the RAW dependency vectors starting from the origin "O" within the base tile, the y-element tiling vector T_y and the x-element tiling vector T_x should satisfy Eqs. (1) and (2), respectively:

$$T_y \geq d_y \max \tag{1}$$

$$
\begin{aligned}
T_x &\geq |OB| - |AB| \\
&= d_x \max - d_x \max \tan \beta \cot \alpha \\
&= d_x \max \cdot (1 - \cot \alpha \tan \beta)
\end{aligned}
\tag{2}
$$

Therefore, the base tile size can be solved as Eq. (3):

$$
\begin{cases}
T_y^{\text{base}} = d_y \max \\
T_x^{\text{base}} = [d_x \max \cdot (1 - \cot \alpha \tan \beta)]
\end{cases}
\tag{3}
$$

For ease of analysis, we propose a *virtual dependency vector* $dv = (d_y v, d_x v)$ to represent the vector which has both the longest y-element length and the longest x-element length of all concerned RAW dependency vectors in an iteration as indicated by the red dashed line in Fig. 7c. The virtual dependency vector is defined as starting from the origin of the base tile and ending at the diagonal. It can be expressed as $d_y = (d_y \max, d_x \max (1 \cot \alpha \tan \beta) + d_y \max \cot \alpha)$, relative to the origin of the base tile. This virtual dependency vector is corresponding to the *virtual MWI* in loop nests. The virtual MWI is the longest define-use chain which is written at the origin and read at the diagonal of the base tile. It covers the longest distance that the real dependency vectors could possibly reach in all dependency directions. So it is convenient and safe to adopt the virtual MWI to represent all the real dependency vectors. The base tile and the virtual MWI offer us a basic framework to derive the optimal tile size.

We call the tile being executed as *current* tile and the one that will be executed next as *next* tile relative to the current one. Any other tiles excluding them are called *remote* tiles relative to the current one. A tile can be partitioned into four parts A, B, C, and D as shown in Fig. 8. The partitioning metrics are as follows:

- All the virtual MWIs written in part A can realize their reuses in the current tile.
- All the virtual MWIs written in part B can only realize their reuses in the next tile.
- All the virtual MWIs written in part C and D can only realize their reuses in the remote tiles.

The parameters and their descriptions with respect to a two-level loop tiling are shown in Table 2.

In terms of part A, all the virtual MWIs written in this area realize their reuses in the current tile. In order to have the MWIs in this area written in fast mode, Eq. (4) should be satisfied:

$$\text{ET}_{(T_y^{\text{base}} \times T_x)} < t_r \tag{4}$$

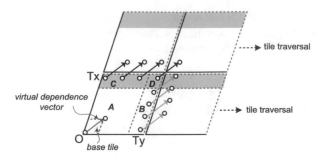

Fig. 8 Tile size analysis

Table 2 Parameter description

Parameter	Description
Size	The whole iteration space of a two-level nested loop
T_y, T_x	Tiling vectors in the y-element and x-element directions, respectively, and x-element directions, respectively
$T_y^{\text{bound}}, T_x^{\text{bound}}$	Upper bounds of tiling vectors in the y-element and x-element directions, respectively
$T_y^{\text{base}}, T_x^{\text{base}}$	y-element vector and x-element vector of the base tile
M	Tile number of the whole loop nests, representing as $M = \text{size}/(T_y-T_x)$
t_r, T_x	The retention time of the MLC PCM with a certain iteration-based programming scheme
t_r^-	Time slot which is close to t_r, representing as $t_r^- = t_r \times (\text{BND} - 1)/\text{BND}$ where BND denotes an integer
$\text{ET}_{(\text{size})}$	The execution time of the iteration space (size)

In terms of part B, all the virtual MWIs written in this area realize their reuses in the next tile. In order to have the MWIs in this area written in the fast mode, Eq. (5) should be satisfied:

$$\text{ET}_{\left(T_y \times T_x + T_y^{\text{base}} \times T_x\right)} < t_r \tag{5}$$

In terms of part C, all the virtual MWIs written in this area realize their reuses in the remote tile. Their lifetime is prolonged by $\text{ET}_{\left((T_y - T_y^{\text{base}}) \times (T_y^{\text{bound}} - T_x)\right)}$. In terms of part D, all the virtual MWIs written in this area realize their reuses in the remote tile. Their lifetime is prolonged by $\text{ET}_{\left((T_y - T_y^{\text{base}}) \times T^{\text{bound}}\right)}$. If the virtual MWIs in parts C and D are written in slow mode in the original loop form by CDDW scheme, it can be seen that these slow writes cannot be transformed into fast mode via loop tiling. In order to have the area of parts C and D as small as possible, it is expected that the value of $T_y^{\text{base}} \times T_x \times M$ is minimized as shown in Eq. (6):

$$\text{Min}\left(T_y^{\text{base}} \times T_x \times M\right) = \text{Min}\left(T_y^{\text{base}} \times T_x \times \frac{\text{size}}{T_y \times T_x}\right)$$
$$\propto \text{Min} \frac{\text{size}}{T_y} \qquad (6)$$

It is observed that Eq. (4) can be merged into Eq. (5). In addition, it is further expected that the tile number M is as small as possible so as to minimize the cost of executing the outer controlling loops in the tiled loop form. This expectation has lower priority than Eq. (6). Putting all together, we can obtain the set of formulas as follows. This formula set covers all the expectations on the optimal size.

$$\begin{cases} \text{(I) } T_x^{\text{base}} \le T_x \le T_x^{\text{bound}} \\ \text{(II) } T_y^{\text{base}} \le T_y \le T_y^{\text{bound}} \\ \text{(III) } \text{ET}_{\left(T_y + T_y^{\text{base}}\right)} \times T_x \Big) < t_r \\ \text{(IV) Min } \frac{\text{size}}{T_y} \text{ (high priority)} \\ \text{(V) Min}(M) = \text{Min}\left(\frac{\text{size}}{T_y \times T_x}\right) \quad \text{(low priority)} \end{cases} \qquad (7)$$

In Formula (7), the subformula set consisting of (I), (II), and (III) aims to obtain the tiling solutions by which the original slow writes in parts A and B can be transformed to fast writes. The subformulas (IV) and (V) further aim to make parts C and D as well as the number of tiles as small as possible.

For loops with iterative regular data array computations, it is reasonable to assume that a MWI's lifetime in a tile is proportional to the MWI's lifetime in the loop nests by the ratio of tile size to the entire loop nest size. The rationale of this assumption is that the execution time of the loop nests is proportional to the loop size. This relationship can be represented by Eq. (8).

$$\frac{\text{ET}_{(\text{size})}}{\text{ET}_{(T_y \times T_x)}} = \frac{\text{size}}{T_y \times T_x} \qquad (8)$$

Applying the deduction in Eq. (8) into the third statement of Formula (7), we can obtain Formulas (9) and (10):

$$T_x < \frac{\text{size} \times t_r}{\text{ET}_{(\text{size})} \times \left(T_y + T_y^{\text{base}}\right)} \qquad (9)$$

$$T_y < \frac{\text{size} \times t_r}{\text{ET}_{(\text{size})} \times T_x} - T_y^{\text{base}} \qquad (10)$$

Since $\text{ET}_{(\text{size})}$ is simulated under fast mode while MWIs in parts C and D should be written in slow mode, a key parameter t_r^- is introduced in our method. It is represented as $t_r^- = t_r \times (\text{BND} - 1)/\text{BND}$ where BND denotes an integer, indicating a time slot close to t_r. This mechanism provides a safe distance between the estimated lifetime and the retention time. The smaller the value of B is, the safer our estimation is. Although $\text{ET}_{(T_y \times T_x)}$ in Eq. (8) is estimated by simulation, we can

184 K. Qiu et al.

Fig. 9 Three cases of loop
tile size selection: (**a**)
$\delta > T_x^{bound}$, (**b**)
$T_x^{base} < \delta \le T_x^{bound}$, (**c**)
$\delta \le T_x^{base}$

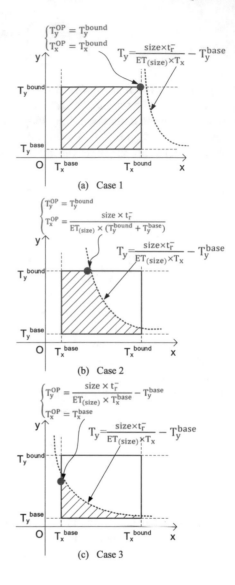

(a) Case 1

(b) Case 2

(c) Case 3

adjust the value of BND during iterative profiling to obtain conservative boundary
of t_r^- sizes to ensure data reliability in parts A and B.

Considering the constraints on T_y and T_x indicated in Formulas (7) (I) and (II)
and Formula (9), there exist three cases to solve the tile size as shown in Fig. 9. We
assign $\delta = \dfrac{size \times t_r^-}{ET_{(size)} \times \left(T_y^{bound} + T_y^{base}\right)}$. Obviously, $\delta > 0$.

Case 1 If $\delta > T_x^{bound}$ as indicated in Fig. 9a, then the solutions in the shaded area
can satisfy Formulas (7)(I), (II), and (III). Considering Formulas (7)(IV) and (V),

we can obtain the optimal tiling solution in *Case 1* as follows. This solution implies that no tiling is needed.

$$
\begin{cases}
T_y^{\text{OP}} = T_y^{\text{bound}} \\
T_x^{\text{OP}} = T_x^{\text{bound}}
\end{cases}
\tag{11}
$$

Case 2 If $T_x^{\text{base}} < \delta \leq T_x^{\text{bound}}$ as indicated in Fig. 9b, then the solutions in the shaded area can satisfy Formulas (7)(I), (II), and (III). Targeting the objectives in Formulas (7)(IV) and (V), we choose $T_y = T_y^{\text{bound}}$ in order to achieve minimal $\frac{\text{size}}{T_y}$ and choose $T_x = \dfrac{\text{size} t_r^-}{\text{ET}_{(\text{size})} \times \left(T_y^{\text{bound}} + T_y^{\text{base}} \right)}$ in order to achieve the minimal tile number. In summary, the optimal tiling solution in Case 2 is as follows:

$$
\begin{cases}
T_y^{\text{OP}} = T_y^{\text{bound}} \\
T_x^{\text{OP}} = \dfrac{\text{size} \times t_r^-}{\text{ET}_{(\text{size})} \times \left(T_y^{\text{bound}} + T_y^{\text{base}} \right)}
\end{cases}
\tag{12}
$$

Case 3 If $\delta \leq T_x^{\text{base}}$ as indicated in Fig. 9c, then the solutions in the shaded area can satisfy Formulas (7)(I), (II), and (III). Targeting the objective in Formula (7)(IV), we choose $T_x = T_x^{\text{base}}$ in order to achieve minimal $\frac{\text{size}}{T_y}$; thus we have $T_y = \dfrac{\text{size} \times t_r^-}{\text{ET}_{(\text{size})} \times T_x^{\text{base}} - T_y^{\text{base}}}$. In summary, the optimal tiling solution in *Case 3* is as follows:

$$
\begin{cases}
T_y^{\text{OP}} = \dfrac{\text{size} \times t_r^-}{\text{ET}_{(\text{size})} \times T_x^{\text{base}}} - T_y^{\text{base}} \\
T_x^{\text{OP}} = T_x^{\text{base}}
\end{cases}
\tag{13}
$$

The worst-case execution time $\text{ET}_{(\text{size})}$ under the fast mode for the entire loop nests can be obtained by simulation. Then the parameter B can be safely worked out by iterative profiling along with the simulation. Thereby, the item δ can be obtained. Finally, we identify the case where the loop nests can be in and calculate the corresponding optimal tile size. The shading area in Fig. 9 covers all the solutions that meet subformulas (I), (II), and (III), and the optimal one indicated by a dot means that it meets all the five subformulas in Formula (7) for each case. The optimal loop tiling result for a given loop indicates the solution that meets all the requirements in Formula (7). In other words, the optimal tiling solution can achieve the maximal fast writes for the MWIs in the loop and meanwhile produce the minimal tile number. Therefore, the "optimal" solution implies the largest benefit on performance improvement and energy saving.

Taking the loop nests of *wave* benchmark as an example, Fig. 10a depicts its source code. The base tile can be obtained through Eq. (3) as shown below:

$$
\begin{cases}
T_y^{\text{base}} = 1 \\
T_x^{\text{base}} = 2
\end{cases}
$$

Fig. 10 Loop tiling for the *wave* kernel. (**a**) Loop kernel of wave. (**b**) Optimal loop tiling

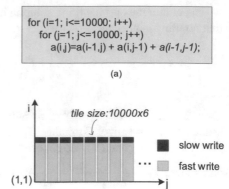

The retention time t_r is *1.87 s* for a fast PCM write. It is further measured that the worst-case execution time under the fast write scheme is 3.3×10^9 cycles. The value of δ can be obtained as 5.67, which indicates that the optimal tile size calculation is involved in Case 2. We specify t_r^- as *1.85 s* by assigning a safe value of BND as slow write fast write 100 through initial profiling. Therefore, we can employ Eq. (12) to work out the optimal tile size as below:

$$\begin{cases} T_y^{OP} = 10000 \\ T_x^{OP} = 6 \end{cases}$$

For the data which are determined to be written in fast mode during loop execution, if they will be read again at some time outside of the loop, they may be unreliable at the time of being reused. In this case, these data can be first identified by liveness analysis. And then rewrite action with slow mode will be taken for them when exiting the loop. Thus, their correctness can be ensured when being reused outside of loop. In our experiments, we found only a few data in the benchmarks have such feature, while most data are just temporarily used inside loop.

4.4 Multilevel Loop Tiling

For multilevel loop nests with multidimension dependence vectors $d = (d_n, d_{n-1}, \ldots, d_1)$, similarly, the legal tiling shape can be derived as follows. The d_1 vector is chosen to be aligned with the positive *1*-axis, and the $d_n, d_{n-1}, \ldots, d_2$ vectors are chosen to be aligned with their CCW vector directions, respectively.

An iterative tiling procedure is used to determine the tiling size. In the first iteration, the *n*-level loop nests are regarded as two-level loop nests consisting of the

innermost first loop and the outer $n-1$ loops. We assign $\delta_1 = \dfrac{size^n \times t_r^{n-}}{ET_{(size^n)} \times (T_{2 \to n}^{bound} + T_{2 \to n}^{base})}$
where $size^n$ denotes the whole iteration space and $T_{2 \to n}$-like expressions denote the tile size starting from the *second* loop to the n-th loop. Following the optimal tile size derivation in Sect. 5.3, there exist three cases to solve the tile size as follows.

Case 1 If $\delta_1 > T_1^{bound}$, the optimal tile size is solved as shown in Eq. (15). This solution implies that no tiling is needed.

$$\begin{cases} T_{2 \to n}^{OP} = T_{2 \to n}^{bound} \\ T_1^{OP} = T_1^{bound} \end{cases} \tag{14}$$

The expression $T_{2 \to n}^{OP} = T_{2 \to n}^{bound}$ can be further finalized by assigning the upper-bound value for each loop. Equation (14) can be transformed into the expressions as below.

$$\begin{cases} T_n^{OP} = T_n^{bound} \\ T_{n-1}^{OP} = T_{n-1}^{bound} \\ \quad \cdots\cdots \\ T_2^{OP} = T_2^{bound} \\ T_1^{OP} = T_1^{bound} \end{cases} \tag{15}$$

Case 2 If $T_1^{base} < \delta_1 \leq T_1^{bound}$, the optimal tile size is solved as shown in Eq. (16).

$$\begin{cases} T_{2 \to n}^{OP} = T_{2 \to n}^{bound} \\ T_1^{OP} = \dfrac{size^n \times t_r^{n-}}{ET_{(size^n)} \times (T_{2 \to n}^{bound} + T_{2 \to n}^{base})} \end{cases} \tag{16}$$

The above equation set can be further finalized as below.

$$\begin{cases} T_n^{OP} = T_n^{bound} \\ T_{n-1}^{OP} = T_{n-1}^{bound} \\ \quad \cdots\cdots \\ T_2^{OP} = T_2^{bound} \\ T_1^{OP} = \dfrac{size^n \times t_r^{n-}}{ET_{(size^n)} \times (T_n^{bound} \times \cdots \times T_2^{bound} + T_n^{base} \times \cdots \times T_2^{base})} \end{cases} \tag{17}$$

Case 3 If $\delta_1 \leq T_1^{base}$, the optimal tile size is solved as shown in Eq. (18).

$$\begin{cases} T_{2 \to n}^{OP} = \dfrac{size^n \times t_r^{n-}}{ET_{(size^n)} \times T_1^{base}} - T_{2 \to n}^{base} \\ T_1^{OP} = T_1^{base} \end{cases} \tag{18}$$

In Cases 1 and 2, the final optimal tile size for every level loop can be obtained directly. In Case 3, only the optimal tile size of the innermost loop and the overall tile size of the outer $n - 1$ loop (i.e., $T^{OP}_{2 \to n}$) can be obtained. Then, we need to further solve the tiling solution for the outer $n - 1$ loops. Analogously, the outer $n - 1$ loop nests can be regarded as two-level loop nests consisting of the innermost *second* loop and the outer $n - 2$ loops. Also, we have $t_r^{n-1} = \frac{t_r^n}{T_1^{base}}$, $\delta_2 = \frac{size^{n-1} \times t_r^{(n-1)-}}{ET_{(size^{n-1})} \times (T^{bound}_{3 \to n} + T^{base}_{3 \to n})}$, and three cases for the outer $(n - 2)$-level loop tiling. This iterative tiling procedure is depicted in Algorithm 1.

Algorithm 1 Iterative loop tiling.

Require:
$size^n$: the whole iteration space
$size^{n-i+1}$: the iteration space from the outermost loop to the ith loop
t_r^n: the retention time for the whole iteration space
t_r^{n-i+1}: the regarded retention time for the iteration space from the outermost loop to the ith loop
$T_{i \to j}$: the tile size starting from the ith loop to the jth loop

1:   $\delta_1 = \frac{size^n \times t_r^{n-}}{ET_{(size^n)} \times (T^{bound}_{2 \to n} + T^{base}_{2 \to n})}$;

2:   **for** $i = 1$ to n **do**

3:     **if** $\delta_i > T_i^{bound}$ **then**

4:       //The final optimal tiling solution from the outermost loop to the ith loop is:

5:       $T^{OP}_{i+1 \to n} = T^{bound}_{i+1 \to n}$;

6:       $T^{OP}_i = T_i^{bound}$;

7:       $i = n + 1$; The overall solution is obtained. Break.

8:     **else**

9:       **if** $T_i^{base} < \delta_i \le T_i^{bound}$ **then**

10:        //The final optimal tiling solution from the outermost loop to the ith loop is:

11:        $T^{OP}_{i+1 \to n} = T^{bound}_{i+1 \to n}$;

12:        $T^{OP}_i = \frac{size^{n-i+1} \times t_r^{(n-i+1)-}}{ET_{(size^{n-i+1})} \times (T^{bound}_{i \to n} + T^{base}_{i \to n})}$

13:        $i = n + 1$; //The overall solution is obtained. Break.

14:       **else**

15:        //$\delta_i \le T_i^{base}$;

16:        //The optimal tiling solution is:

17:        $T^{OP}_{i \to n} = \frac{size^{n-i+1} \times t_r^{(n-i+1)-}}{ET_{(size^{n-i+1})} \times T_i^{base}} - T^{base}_{n-i+1 \to n}$;

18:        $T^{OP}_i = T_i^{base}$;

19:        //Recalculate δ and the regarded retention time for the next round

20:        $\delta_{i+1} = \frac{size^{n-i} \times t_r^{(n-i)-}}{ET_{(size^{n-i})} \times (T^{bound}_{i+2 \to n} + T^{base}_{i+2 \to n})}$;

21:        $t_r^{n-i} = \frac{t_r^n}{T_{1 \to i}^{base}}$;

22:        i++;

23:       **end if**

24:     **end if**

25:   **end for**

26: **return** *the final optimal tiling results*;

5   Experiments

In this section, we first introduce the experimental setup. Then the results on performance and dynamic energy by applying the proposed approach are presented. Furthermore, the sensitivity to PCM size is discussed. Finally, we discuss the accuracy of the proposed deductions.

5.1   Experimental Setup

This work targets energy efficient MCUs which have on-chip or off-chip memory built by PCM, but do not have caches. MCUs with similar targeting configuration include Tiny12 [28], ARM7TDMI [29], megaAVR [30], and PIC [31] series. These MCUs have neither cache nor SPM. Therefore, CPU can access the off-chip main memory directly. There are also MCUs that adopt an on-chip nonvolatile memory. For example, MSP430FR series [32] have on-chip FRAM memory and SRAM memory that work as SPM. In both cases, the nonvolatile memories are part of the main memory address space, and CPUs can directly access them. Therefore, the proposed loop tiling approach in our work can be applied. In the experiments, we choose the 10-iteration programming mode as the slow mode and the 3-iteration programming mode as the fast mode. The baseline configuration is illustrated in Table 3.

It is assumed that the PCM size equals to the real size utilized by the programs. To evaluate the proposed approach, totally four schemes are evaluated, as depicted in Table 4.

In the fast scheme, all writes are conducted in fast mode, so the MWIs with lifetime longer than the fast write retention time need refreshes. For the data arrays which are useful during the whole program execution time, a DRAM-style refresh method is employed to refresh them during the whole execution period. For the data arrays whose MWIs' lifetime only lasts for a period of the whole program execution time, a reasonable *N-refresh* method [33] is employed to refresh them, where the

Table 3 Parameter description

Component	Parameter
MCU core	Single issue, 1 MHz, no cache, no MMU
Code memory	1 cycle per instruction without access to data memory
Data memory	32-bit width
	Read cost: 1 μs and 48 pJ
	Fast write cost: 3 μs and 955.2 pJ
	Slow write cost: 10 μs and 1542.4 pJ
	Fast write retention time: 1.87 s
	Slow write retention time: 11158.84 s

Table 4 Write schemes for experimental evaluation

Write scheme	Brief description
Fast	All writes are conducted in fast mode, requiring refreshes
Slow	All writes are conducted in slow mode. No refresh
CDDW	A MWI is conducted in fast mode if and only if all its instances are conservatively estimated to have a short lifetime less than the fast write retention time. No refresh
WMALT + CDDW	The proposed WMALT preprocessing is conducted to effectively reduce MWIs' lifetime in loops before applying the CDDW scheme

data arrays are refreshed $2^N - 1$ times of the fast write retention time. The N-refresh method prolongs the fast MWIs' validation time. It is more performance-efficient and energy-efficient than the DRAM-style refresh method. In the evaluation, the value of N is set to be 5.

In the CDDW scheme, write modes are statically selected for different MWIs [15]. In the WMALT + CDDW scheme, the proposed WMALT approach is embedded to preprocess the loops before implementing the CDDW scheme.

The WCLT analysis is implemented on the Chronos platform [34] using the method [15]. An initial iterative profiling is conducted to obtain t_r^- and ensure that it is safe enough to guarantee MWIs' correctness of parts A and B. For each benchmark, we have four evaluation versions, corresponding to the four write schemes in Table 4. For the WMALT + CDDW scheme, loop tiling is determined based on the Worst-case execution time (WCET) results under the fast schemes. The loop tiling preprocessing is done on the source code. For each version, the benchmarks are compiled by GCC "static." An initial iterative profiling will be conducted to ensure that t_r^- is safe enough to guarantee MWIs' correctness of parts A and B. By analyzing the binary file together with the WCLT information, different write latencies and energy consumptions are assigned to the MWIs; thus the performance and energy results can be obtained.

The benchmarks are all loop kernels, extracted from DSP programs, Blitz++ library [35], and MiBench suite [36]. Basic characteristics of the selected benchmarks are given in Table 5. We assign diverse loop sizes for the loop nests. The optimal tiling size calculated by the proposed WMALT approach is listed in Table 6. In the original loop form, at least one data array's MWIs are characteristic of their WCLT longer than the fast retention time.

5.2 Results and Analysis

In this subsection, the four schemes are compared in terms of performance and dynamic energy consumption.

Table 5 Benchmark characteristics

Benchmark	Description	Dynamic instructions	Reads	Writes
Wave	Wavefront computation	4.4E + 09	1.4E + 09	2.1E + 08
wdf	Wave digital filter	3.8E + 09	1.2E + 09	2.1E + 08
iir	Infinite impulse response	1.1E + 10	3.8E + 09	2.1E + 08
acou2d	Acoustic2d	1.7E + 10	4.0E + 09	4.0E + 08
acou3d	Acoustic3d	3.3E + 10	6.4E + 10	4.8E + 09
Stencil	Array stencil	3.3E + 10	6.4E + 10	4.8E + 09
Array	Array computation	3.3E + 10	6.4E + 10	4.8E + 09
formatBit	formatBitstream module in Lame	9.9E + 08	3.3E + 08	1.6E + 08
Floyed	Floyed algorithm	4.8E + 10	8.0E + 09	2.0E + 09
mm	Matrix multiplication	2.8E + 10	9.5E + 09	2.9E + 09

Table 6 Data array size and optimal loop tiling size

Benchmark	Loop size	Array #	Optimal tiling size
Wave	$10^4 \times 10^4$	3	$10{,}000 \times 6$
wdf	$10^4 \times 10^4$	1	8250×2
iir	$10^4 \times 10^4$	3	331×4
acou2d	$10^4 \times 10^4$	4	$10{,}000 \times 2$
acou3d	$10^3 \times 10^3 \times 10^3$	4	$1000 \times 1000 \times 16$
Stencil	$10^3 \times 10^3 \times 10^3$	2	$1000 \times 1000 \times 15$
Array	$10^3 \times 10^3 \times 10^3$	2	$1000 \times 1000 \times 11$
formatBit	$10^4 \times 10^4$	3	$10{,}000 \times 3$
Floyed	$10^3 \times 10^3 \times 10^3$	1	$1000 \times 343 \times 2$
mm	$10^3 \times 10^3$	3	$1000 \times 284 \times 2$

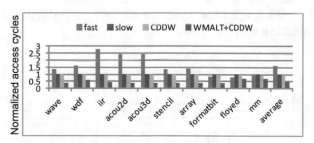

Fig. 11 Performance improvement

5.2.1 Performance

Figure 11 shows the performance comparisons among the four write schemes. The results are normalized to the slow scheme. We have the following observations:

For most benchmarks, the fast scheme achieves the worst performance, which is caused by the expensive refresh cost. For other benchmarks, such as *formatbit, floyed*, and *mm*, the fast scheme has better performance than that under the slow scheme. The reason lies in that, the involved data array number and the refresh number are both small; thus the fast scheme works better with small refresh cost.

For all benchmarks, the performance of the pure CDDW scheme is close to that of the slow scheme. This is because most MWIs' lifetime is larger than the fast write retention time, being assigned slow mode. The proposed WMALT + CDDW scheme always achieves the best performance. On average, it outperforms the fast, slow, and CDDW schemes by 70.0%, 51.8%, and 50.8%, respectively. Compared to the fast scheme with large refresh cost and the slow scheme with long write latency, the WMALT approach can effectively reduce most of MWIs' lifetime so that most write instances can be written in fast mode. The CDDW scheme guarantees that each write instance is written in the more efficient mode. Combining these two approaches, the performance is significantly improved.

5.2.2 Dynamic Energy

The dynamic energy includes the dynamic write and read energy. Figure 12 shows the dynamic energy under the four schemes. All results are normalized to the slow scheme. The fast scheme costs the most dynamic energy for most benchmarks. The WMALT + CDDW scheme reduces dynamic energy by 35.6%, 33.7%, and 32.0% compared to the fast, slow, and CDDW schemes, respectively.

For *acou2d, acou3d*, and *iir*, the improvement by the proposed approach is significantly compared to the fast scheme. The reason is that, the number of refresh operations is large relative to the number of writes in these benchmarks, and thus the refresh overhead is significant. This also explains the phenomenon that the slow scheme works better than the fast scheme for these three applications. In this case, the WMALT approach can bring large benefit by reducing MWIs' lifetime and enabling maximal fast writes.

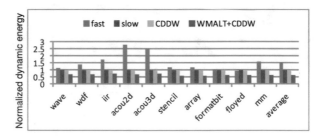

Fig. 12 Dynamic energy improvement

5.3   Sensitivity Study of PCM Size

Furthermore, the effectiveness of the proposed WMALT + CDDW scheme is discussed by changing PCM sizes for measurements. Figures 13 and 14 show the impacts of PCM size on performance and energy consumption for the benchmarks *iir*, *formatbit*, and *stencil*, respectively. The PCM size is represented by the loop size as shown in Table 7.

We have the following observations:

As the loop size becomes larger, the fast scheme degrades sharply because the large additional cost from refresh operations is introduced. This refresh cost is proportional to the data array size. In this case, the CDDW scheme chooses slow mode for the MWIs of data arrays, presenting similar performance to the slow scheme. However, the WMALT approach can make most MWIs' lifetime reduced

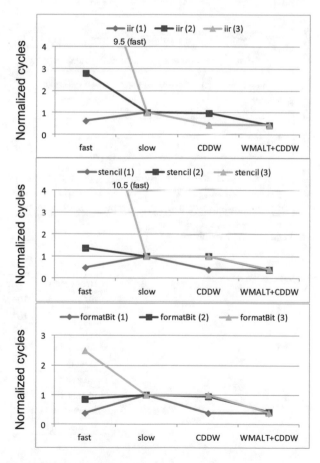

Fig. 13 Impacts of PCM size on performance

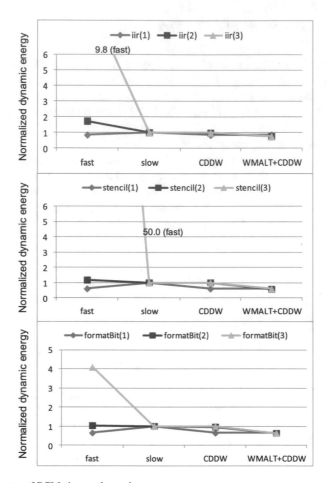

Fig. 14 Impacts of PCM size on dynamic energy

Table 7 Variable loop size

Benchmark	Loop size
iir(1)	$10^3 \times 10^3$
iir(2)	$10^4 \times 10^4$
iir(3)	$10^5 \times 10^5$
stencil(1)	$10^2 \times 10^2 \times 10^2$
stencil(2)	$10^3 \times 10^3 \times 10^3$
stencil(3)	$10^4 \times 10^4 \times 10^4$
formatBit(1)	$10^3 \times 10^3$
formatBit(2)	$10^4 \times 10^4$
formatBit(3)	$10^5 \times 10^5$

below the fast write retention time bound. Thus the WMALT + CDDW scheme can make large possible MWIs written in fast mode to achieve significant improvement on performance and dynamic energy.

As the loop size becomes smaller, the MWIs' lifetime of data arrays becomes smaller than the PCM fast write retention time. Thus most write operations in the applications can be written in fast mode with no refreshes under the CDDW scheme to achieve high performance and low power. There is no need to call the WMALT approach.

In either case, the WMALT + CDDW scheme always offers double guarantees to outperform the slow scheme and the fast schemes and achieves good improvement.

5.4 Discussion on Accuracy

Addressing data reliability under fast write mode, we have taken multiple steps to obtain conservative loop tiling solutions to ensure data correctness. First, the retention time of the volatile MLC PCM model is generated taking into account the worst cases of process variation and resistance drifting ratio. In fact, the retention time t_r in Eqs. 9 and 10 has been underestimated. Second, the write mode selection is determined based on static analysis of write instructions' worst-case lifetime (WCLT) [15]. This WCLT analysis has overestimated the lifetime of each write instruction. Third, the parameter BND provides a safe distance between the estimated lifetime and the retention time. It is assigned to first ensure that MWIs' lifetime in area B (as well as area A) does not exceed the retention time of fast write mode of MLC PCM. Meanwhile, the parameter BND should make the introduced parameter t_l^- as close as possible to the retention time t_r to obtain as large as possible performance and power improvements. Altogether, all the above mechanisms can largely provide a conservative calculation and try to make sure that MWIs' lifetime in area B is less than the retention time in fast mode with our tiling method.

This work focuses on stencil loops in embedded applications such as digital signal processing, numerical simulations, and image processing. In these loops, the computation is regular and predictable, where Eq. (8) will hold. Therefore, the proposed technique can always ensure the correctness.

However, if each iteration's execution time is not fixed, the proposed technique does not always apply. In this case, the estimation of execution time of a loop tile is not accurate. The lifetime of some elements may exceed the retention time of fast write mode and lead to data unreliability. We will investigate this scenario and propose techniques to handle this situation in our future work.

6 Conclusion

This work presents a write mode aware loop tiling (WMALT) approach to effectively reduce memory write instances' lifetime and maximize fast writes in stencil loop applications. Hence, the performance and energy consumption for loops on MLC PCM can be significantly improved. The optimal tile size determination is fully discussed. Experimental validation demonstrates the effectiveness of the WMALT approach.

Acknowledgments This work is supported by grants from Beijing Advanced Innovation Center for Imaging Technology, Beijing Municipal Innovative Science and Technology Platform, the Project of Construction of Innovative Teams and Teacher Career Development for Universities and Colleges Under Beijing Municipality [Project No. IDHT20150507], National Natural Science Foundation of China [Project No. 61472260, 61402302, 61502321, and 61872251], and Beijing Natural Science Foundation of China [Project No. 4132016 and 4143060].

References

1. Zhou P, Zhao B, Yang J, Zhang Y (2009) A durable and energy efficient main memory using phase change memory technology. In: Proceedings of the 36th annual international symposium on computer architecture (ISCA), pp 14–23
2. Qureshi MK, Srinivasan V, Rivers JA (2009) Scalable high performance main memory system using phase-change memory technology. In: Proceedings of the 36th annual international symposium on computer architecture (ISCA), pp 24–33
3. Raoux S, Burr G, Breitwisch M, Rettner C, Chen Y, Shelby R, Salinga M, Krebs D, Chen S-H, Lung HL, Lam C (2008) Phase-change random access memory: a scalable technology. IBM J Res Dev 52(4.5):465–479
4. Xue CJ, Zhang Y, Chen Y, Sun G, Yang JJ, Li H (2011) Emerging non-volatile memories: opportunities and challenges. In: Proceedings of the seventh IEEE/ACM/IFIP international conference on hardware/software code-sign and system synthesis (CODES + ISSS), pp 325–334
5. Jiang L, Zhang Y, Childers BR, Yang J (2012) FPB: fine-grained power budgeting to improve write throughput of multi-level cell phase change memory. In: IEEE/ACM international symposium on microarchitecture (MICRO), pp 1–12
6. Jiang L, Zhao B, Zhang Y, Yang J, Childers B (2012) Improving write operations in MLC phase change memory. In: 2012 IEEE 18th international symposium on high performance computer architecture (HPCA), pp 1–10
7. Qureshi M, Franceschini M, Jagmohan A, Lastras L (2012) PreSET: improving performance of phase change memories by exploiting asymmetry in write times. In: 39th annual international symposium on computer architecture (ISCA), pp 380–391
8. Bedeschi F, Fackenthal R, Resta C, Donze E, Jagasivamani M, Buda E, Pellizzer F, Chow D, Cabrini A, Calvi G, Faravelli R, Fantini A, Torelli G, Mills D, Gastaldi R, Casagrande G (2009) A bipolar-selected phase change memory featuring multi-level cell storage. IEEE J Solid State Circuits (JSSC) 44(1):217–227
9. Nirschl T, Philipp J, Happ T, Burr G, Rajendran B, Lee MH, Schrott A, Yang M, Breitwisch M, Chen C, Joseph E, Lamorey M, Cheek R, Chen SH, Zaidi S, Raoux S, Chen Y, Zhu Y, Bergmann R, Lung HL, Lam C (2007) Write strategies for 2 and 4-bit multi-level phase-change memory. In: IEEE international electron devices meeting (IEDM), pp 461–464

10. Hay A, Strauss K, Sherwood T, Loh G, Burger D (2011) Preventing PCM banks from seizing too much power. In: IEEE/ACM international symposium on microarchitecture (MICRO), pp 186–195
11. Hu J, Zhuge Q, Xue CJ, Tseng WC, Sha EHM (2013) Software enabled wear-leveling for hybrid pcm main memory on embedded systems. In: Design, automation & test in Europe conference & exhibition (DATE), pp 599–602
12. Liu T, Zhao Y, Xue CJ, Li M (2011) Power-aware variable partitioning for DSPs with hybrid PRAM and DRAM main memory. In: 48th ACM/EDAC/IEEE design automation conference (DAC), pp 405–410
13. Awasthi M, Shevgoor M, Sudan K, Rajendran B, Balasubramonian R, Srinivasan V (2012) Efficient scrub mechanisms for error-prone emerging memories. In: IEEE 18th international symposium on high performance computer architecture (HPCA), pp 1–12
14. Lin JT, Liao YB, Chiang MH, Chiu IH, Lin CL, Hsu WC, Chiang PC, Sheu SS, Hsu YY, Liu WH, Su KL, Kao MJ, Tsai MJ (2009) Design optimization in write speed of multi-level cell application for phase change memory. In: IEEE international conference of electron devices and solid-state circuits (EDSSC), pp 525–528
15. Li Q, Jiang L, Zhang Y, He Y, Xue CJ (2013) Compiler directed write-mode selection for high performance low power volatile PCM. In: Proceedings of the 14th ACM SIGPLAN/SIGBED conference on languages, compilers and tools for embedded systems (LCTES), pp 101–110
16. Li Z, Song Y (2004) Automatic tiling of iterative stencil loops. ACM Trans Program Lang Syst 26(6):975–1028
17. Xue J (2000) Loop tiling for parallelism
18. Di P, Wu H, Xue J, Wang F, Yang C (2012) Parallelizing SOR for GPGPUs using alternate loop tiling. Parallel Comput 38(6–7):310–328
19. Wolf M, Lam M (1991) A loop transformation theory and an algorithm to maximize parallelism. IEEE Trans Parallel Distrib Syst 2(4):452–471
20. Renganarayanan L, Kim D, Strout MM, Rajopadhye S (2012) Parameterized loop tiling. ACM Trans Program Lang Syst 34(1): 3: 1–3: 41
21. Borkar S, Karnik T, Narendra S, Tschanz J, Keshavarzi A, De V (2003) Parameter variations and impact on circuits and microarchitecture. In: Proceedings of the 40th annual design automation conference (DAC), pp 338–342
22. Zhang W, Li T (2009) Characterizing and mitigating the impact of process variations on phase change based memory systems. In: Proceedings of the 42nd annual IEEE/ACM international symposium on microarchitecture (MICRO), pp 2–13
23. Dong J, Zhang L, Han Y, Wang Y, Li X (2011) Wear rate leveling: lifetime enhancement of PRAM with endurance variation. In: Proceedings of the 48th design automation conference (DAC), pp 972–977
24. Jiang L, Zhang Y, Yang J (2011) Enhancing phase change memory lifetime through fine-grained current regulation and voltage upscaling. In: Proceedings of the 17th IEEE/ACM international symposium on low-power electronics and design (ISLPED), pp 127–132
25. Zhang W, Li T (2011) Helmet: a resistance drift resilient architecture for multi-level cell phase change memory system. In: IEEE/IFIP 41st international conference on dependable systems networks (DSN), pp 197–208
26. Jung CM, Lee ES, Min KS, Kang SMS (2011) Compact verilog-A model of phase-change RAM transient behaviors for multi-level applications. In: Semiconductor Science and Technology, vol 25(7)
27. Chen F, O'Neil T, Sha E-M (2000) Optimizing overall loop schedules using prefetching and partitioning. IEEE Trans Parallel Distrib Syst 11(6):604–614
28. ATTiny12. www.atmel.com/devices/attiny12.aspx/
29. ARM7TDMI. http://infocenter.arm.com/help/topic/com.arm.doc.ddi0210c/DDI0210B.pdf
30. megaAVR. www.atmel.com/products/microcontrollers/avr/megaAVR.aspx
31. PIC. www.microchip.com/pagehandler/en-us/products/
32. MSP430FR.   http://www.ti.com/lsds/ti/microcontrollers16-bit32-bit/msp/ultra-lowpower/msp430frxxfram/overview.page

33. Li Q, Li J, Shi L, Xue C, Chen Y, He Y (2013) Compiler-assisted refresh minimization for volatile STT-RAM cache. In: 2013 18th Asia and South Pacific design automation conference (ASP-DAC), pp 273–278
34. Li X, Liang Y, Mitra T, Roychoudury A (2007) Chronos: a timing analyzer for embedded software. Sci Comput Program 69(1–3):56–67
35. Blitz++. http://blitzplus-pplus-p.sourcearchive.com/
36. Guthaus M, Ringenberg J, Ernst D, Austin T, Mudge T, Brown R (2001) MiBench: a free, commercially representative embedded benchmark suite. In: IEEE international workshop on workload characterization, pp 3–14

Index

© Springer Nature Switzerland AG 2020
Y. Liu et al. (eds.), *Smart Sensors and Systems*,
https://doi.org/10.1007/978-3-030-42234-9

Printed in the United States
by Baker & Taylor Publisher Services